QGIS and Applications in Agriculture and Forest

QGIS in Remote Sensing Set

coordinated by
André Mariotti

Volume 2

QGIS and Applications in Agriculture and Forest

Edited by

Nicolas Baghdadi
Clément Mallet
Mehrez Zribi

WILEY

First published 2018 in Great Britain and the United States by ISTE Ltd and John Wiley & Sons, Inc.

ISTE Ltd
27-37 St George's Road
London SW19 4EU
UK

www.iste.co.uk

John Wiley & Sons, Inc.
111 River Street
Hoboken, NJ 07030
USA

www.wiley.com

Library of Congress Control Number: 2017962204

British Library Cataloguing-in-Publication Data
A CIP record for this book is available from the British Library
ISBN 978-1-78630-188-8

Contents

Chapter 3. Automatic Extraction of Agricultural Parcels from Remote Sensing Images and the RPG Database with QGIS/OTB   77

Jean-Marc GILLIOT, Camille LE PRIOL, Emmanuelle VAUDOUR and Philippe MARTIN

Chapter 4. Land Cover Mapping Using Sentinel-2 Images and the Semi-Automatic Classification Plugin: A Northern Burkina Faso Case Study.   119

Louise LEROUX, Luca CONGEDO, Beatriz BELLÓN, Raffaele GAETANO and Agnès BÉGUÉ

Chapter 5. Detection and Mapping of Clear-Cuts with Optical Satellite Images

Kenji OSE

Chapter 6. Vegetation Cartography from Sentinel-1 Radar Images

Pierre-Louis FRISON and Cédric LARDEUX

Introduction

Agriculture and forestry are fields strongly involved in the use of spatial data, which are essential for monitoring and restoring the spatial and temporal variability of surface states. The latter are key parameters in the understanding and modeling of different plant and soil processes, and in the management of agricultural or forest resources. A very good knowledge of these environments is therefore fundamental both from an economic and ecological point of view. Remote sensing, thanks to the great diversity of spatial (from precision agriculture to global crop monitoring), spectral (active and passive sensors) and temporal (from rapid mapping to annual crop monitoring) resolutions, has become an inevitable support to address these issues. In this context, the use of Geographic Information System (GIS) tools has long been present in accompanying the exploitation of spatial imagery.

The aim of this second volume is to present different applications in agriculture and forestry. The book, which is supported by scientists who are internationally renowned in their fields, will help update knowledge and describe research and development issues for years to come. It is intended for research teams in geomatics, second-cycle students (engineering schools, master's degrees) and postgraduate studies (PhD students), and engineers involved in the monitoring and management of agricultural or forestry resources and more fundamentally in the extraction of the knowledge required for these needs. In addition to the texts of the proposed chapters, readers will have access to the data, computer tools as well as screenshots of all the windows, which illustrate all the steps necessary for the realization of each application.

The first chapter of this volume concerns the estimation of the hydric state of the soil by synergy of radar/optical satellite data. Chapter 2 deals with the disaggregation of thermal data. The third chapter discusses the operational and automatic extraction of agricultural fields from satellite imagery and the French Land Parcel Identification System (RPG). Chapter 4 analyzes an application related to land use mapping. The second part of this volume is devoted to forestry applications and includes five chapters. They cover different applications related to forest mapping with active and passive sensors, in different environments, and clear-cut monitoring.

A supplement to the chapters, including data sources (images, training and validation data, auxiliary information) and screenshots illustrating the practical application of the chapters, is available at the following address:

Using Internet Explorer: ftp://193.49.41.230
Using a FileZilla client: 193.49.41.230

Username: **vol2_en**
Password: **n34eVol@2**

We would like to thank the scientists who contributed to the development of this volume, the authors of the chapters of course, but also the experts of the reading committee. This project was carried out with the support of IRSTEA (French National Research Institute of Science and Technology for Environment and Agriculture), IGN (French National Institute for Geographic and Forestry Information), CNRS (French National Center for Scientific Research) and CNES (French National Center for Space Studies).

We are very grateful to Airbus Defense and Space, CNES and Equipex Geosud for providing us with SPOT-5/6/7 images. Please note that these images may only be used in a research and training framework, and any commercial activity based on the data provided is strictly prohibited.

Our thanks also go to our families for their support, and to André Mariotti (Professor Emeritus, Pierre and Marie Curie University) and Pierrick Givone (President, IRSTEA) for their encouragement and support in the realization of this project.

Nicolas BAGHDADI

Clément MALLET

Mehrez ZRIBI

Coupling Radar and Optical Data for Soil Moisture Retrieval over Agricultural Areas

1.1. Context

The spatio-temporal monitoring of soil moisture in agricultural areas is of great importance for numerous applications, particularly those related to the continental water cycle. The use of *in situ* sensors ensures this monitoring but the technique is very costly and can only be carried out on a very small agricultural area, hence the importance of spatial remote sensing that now enables large-scale operational mapping of soil moisture with high spatio-temporal resolution.

Radar data have long been used to estimate and map the surface soil moisture of bare soils [BAG 16b]. In fact, physical, empirical and semi-empirical models were developed to invert the radar signal to monitor the soil moisture at different spatial scales (intra-plot scale, plot scale, on grids of a few hundred m² to a few km²). Over vegetated cover surfaces, the coupling of radar and optical data is often necessary to estimate the surface soil moisture. Optical data are complementary to radar data, and their interest lies in their potential to estimate the physical parameters of vegetation, for example Leaf Area Index (LAI) from satellite indices such as the Normalized Difference Vegetation Index (NDVI). These parameters make it possible to evaluate the contribution of the vegetation in the backscattered radar signal, to extract the soil contribution and to finally invert it in order to estimate the surface soil moisture.

To map the soil moisture in the case of vegetation cover, most studies use the semi-empirical Water Cloud Model (WCM) developed by Attema and Ulaby [ATT 78]. Generally, in this model, the total backscattered radar signal is modeled

Chapter written by Mohammad EL HAJJ, Nicolas BAGHDADI, Mehrez ZRIBI and Hassan BAZZI.

as the sum of (1) the backscattered signal from the soil multiplied by the two-way attenuation and (2) the direct reflected signal from the vegetation. In most studies, the contribution of vegetation has been expressed in terms of one physical parameter of vegetation (biomass, LAI, water content, vegetation height). The soil contribution is generally modeled as a function of soil moisture and surface roughness (for given instrumental parameters: incidence angle, wavelength and polarization). It can be simulated using a physical radar backscattering model (in particular the Integral Equation Model (IEM) [FUN 94]), or a semi-empirical backscattering model (e.g. Dubois [DUB 95] or Baghdadi [BAG 16a] models).

The objective of this chapter is to show how to map the surface soil moisture over agricultural plots (summer and winter crops) and grasslands using the free and open-source software QGIS (Quantum Geographic Information System), by coupling radar (Synthetic Aperture Radar (SAR)) and optical images acquired at high spatial resolution (~10 m × 10 m).

1.2. Study site and satellite data

The study site located near Montpellier in the South of France (Figure 1.1) is an agricultural area (15 km × 15 km). Figure 1.1 shows the layout, made using QGIS, of a satellite image acquired over the study site by Sentinel-2A (S2A).

QGIS functionality for layout:
- Project > *New Print Composer* >...

1.2.1. *Radar images*

Two Sentinel-1A (S1A) radar images in C-band (radar wavelength ~5.6 cm) acquired on January 19, 2017, and January 26, 2017, have been used. On January 19, 2017, the soils in the study site were dry (no precipitation for 19 days, a soil moisture around 11 vol.% is measured on a reference plot), whereas on January 26 the soils were very wet (a soil moisture around 30 vol.% is measured on a reference plot) due to the high rainfall that occurred over 4 days before the radar image acquisition (accumulation of 23 mm). S1A images are freely available from the Copernicus[1] and Google Earth Engine[2] websites. The Copernicus website offers raw images that require radiometric (passage of digital number into backscattering coefficient) and geometric calibration. The downloaded images from

1 https://scihub.copernicus.eu/dhus/#/home.
2 https://earthengine.google.com.

Google Earth Engine are already calibrated and ortho-rectified (WGS84 projection system).

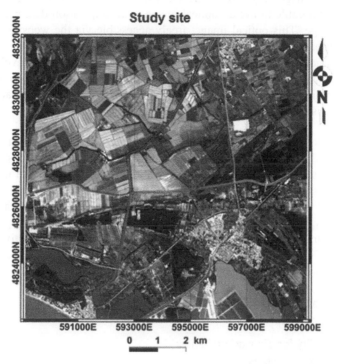

Figure 1.1. *Study site located 5 km east of Montpellier. The background of the map is an optical image acquired by the satellite S2A. The geographical coordinates are in UTM (Universal Transverse Mercator), zone 31 N. For a color version of the figure, see www.iste.co.uk/baghdadi/QGIS2.zip*

Radar images used in this chapter have been downloaded from the Google Earth Engine website. Each image is a stack of three bands: band 1 corresponds to the backscattering coefficient in VV polarization (in decibel (dB) scale), band 2 is the backscattering coefficient in VH polarization (in dB) and band 3 contains the local incidence angle relative to the ellipsoid (in degrees). The two bands corresponding to backscattering coefficients in VV and VH polarizations have been transformed into linear scale. Part 1 of the flowchart (Figure 1.2) shows the processing performed on radar images.

QGIS functionality to transform the first two bands of radar images in linear scale:
- Raster> *Raster Calculator >...*

1.2.2. *Optical image*

One optical image acquired by the satellite S2A on October 15, 2016, is used. Ideally, it is preferable to use an optical image at an acquisition date near to that of each radar image. This optical image, freely accessible via the website of the land data center Theia[3], covers an area of 110 km × 110 km. The Theia website provides S2A data corrected from atmospheric and slope effects (processing level 2A). S2A images are downloadable from the Theia website in the form of 13 separate spectral bands. The projection system associated with the S2A images downloaded via the Theia website is also the UTM.

To facilitate the use of an optical image, three spectral bands in the visible (bands 2, 3 and 4) and one in the infrared domain (band 8) are first stacked. The optical image is then clipped to adjust the spatial extent of the optical image to the surface of the study site (15 km × 15 km). Next, the clipped image is reprojected into a WGS84 geodetic system to be in the same projection system as the radar images. Finally, an NDVI image is calculated from the reprojected optical image using the spectral bands corresponding to the red and infrared (respectively, band 4 and band 8). The second part of the flowchart (Figure 1.2) shows the processing performed on optical image.

QGIS functionality for stacking the four spectral bands:
- Raster > *Miscellaneous > Build Virtual Raster >...*

QGIS functionality for clip stacked bands:
- Raster > *Extraction > Clipper >...*

QGIS functionality to reproject the image:
- Raster > *Projection > Warp (Reproject) >...*

QGIS functionality to calculate the NDVI image:
- Raster > *Raster Calculator >...*

1.2.3. *Land cover map*

A land cover map[4] produced by the scientific expertise center of Theia is used to extract the crop plots and grasslands. This map is a thematic raster file with values between 11 and 222, where each value corresponds to a type of land cover[5]. The projection system associated to the land cover map of Theia is Lambert-93. The land

---

3 https://theia.cnes.fr/atdistrib/rocket/#/search?collection=SENTINEL2.
4 http://osr-cesbio.ups-tlse.fr/echangeswww/TheiaOSO/OCS_2014_CESBIO_L8.tif.
5 http://osr-cesbio.ups-tlse.fr/~oso/ui-ol/2009-2011-v1/layer.html.

cover map is first clipped to adjust the spatial extent of the study site. Next, the clipped map is reprojected in the WGS84 geodesic system to have the same projection system as that of radar and optical images. Part 3 of the flowchart (Figure 1.2) shows the processing done with the land cover map.

1.3. Methodology

In this section, the steps that lead to the production of soil moisture maps on crop areas and grasslands are described. First, an inversion approach using neural networks is developed. The networks are trained using a simulated dataset of radar backscattering coefficients obtained from the WCM. In WCM, the IEM calibrated by Baghdadi *et al.* [BAG 06] is used to simulate the soil contribution. The application of neural networks on real satellite data requires the identification of crop and grassland zones. These zones have been extracted from the land cover map available on the study site. Next, an NDVI image calculated from the optical image is used to partition these zones into homogeneous segments (intra-plot scale). Finally, the soil moisture maps are produced by applying the developed inversion approach on each homogeneous segment.

1.3.1. *Inversion approach of radar signal for estimating soil moisture*

The soil of agricultural areas is covered for a long period of the year by vegetation. An approach that considers the effects of vegetation on the backscattered radar signal for estimating the soil moisture is therefore indispensable for accurate estimation of the soil moisture.

The WCM defines the backscattered radar signal in linear scale (σ^0_{tot}) as the sum of the contribution from the vegetation (σ^0_{veg}), the contribution of soil (σ^0_{soil}) attenuated by the vegetation ($T^2 \sigma^0_{soil}$) and multiple soil–vegetation scatterings (often neglected):

$$\sigma^0_{tot} = \sigma^0_{veg} + T^2 \sigma^0_{soil} \qquad [1.1]$$

$$\sigma^0_{veg} = AV_1 cos\theta (1 - T^2) \qquad [1.2]$$

$$T^2 = e^{-2BV_2 sec\theta} \qquad [1.3]$$

where:

– V_1 and V_2 are the vegetation descriptors: biomass, vegetation water content, vegetation height, LAI, NDVI (in this chapter, $V_1 = V_2 = $ NDVI);

– θ is the radar incidence angle (°);

– A and B are fitting parameters of the model that depend on the chosen vegetation descriptor and the radar configuration.

The soil contribution σ^0_{soil} that depends on soil moisture and surface roughness (in addition to SAR instrumental parameters) is simulated in this chapter using the physical backscattering model IEM, calibrated by Baghdadi *et al.* [BAG 06].

The steps for designing the soil moisture estimation algorithm are as follows:

– Calibrate the WCM using experimental data obtained on reference plots: radar signal, NDVI (from optical images) and *in situ* measurements of soil moisture and surface roughness carried out during the radar sensor overpass. This model calibration phase leads to the calculation of the parameters A and B.

– Generate synthetic data of backscattered radar signals by using the calibrated WCM and the IEM model [BAG 06] for a wide range of soil moisture values (between 2 and 40 vol.%), of surface roughness (between 0.5 and 4.0 cm) and of NDVI (between 0 and 1), in order to cover all possible soil and vegetation parameter values in agricultural contexts. In addition, the radar incidence angle in the WCM is varied according to the range of incidence angle available on the radar sensors (between 20° and 45°).

– Add noise to synthetic data to better simulate an experimental dataset of radar signals (additive Gaussian noise with zero mean and a standard deviation of 1 dB [MIR 15]), and NDVI (relative additive noise of 15% on the NDVI [EL 09]).

– Train the neural networks using synthetic dataset. A detailed description of this procedure is given in [EL 16], as well as a justification of the used method.

In this chapter, we present one of the trained neural networks. For a given radar acquisition date, the radar signals in VV and VH polarizations as well as the local incidence angle and the NDVI values (calculated from an optical image) are the inputs of the neural network. At the output of the neural network, we obtain the soil moisture, estimated in vol.%.

1.3.2. *Segmentation of crop and grasslands areas*

We seek to estimate soil moisture for each homogeneous spatial unit (sub-plot, plot or set of plots), defined by pixels with homogeneous NDVI values with a variation of ±0.1. For this, it is necessary to delineate each homogeneous spatial unit (a polygon). A mask is first generated from the land cover map to extract the crop areas and grasslands. Next, the NDVI image is used to segment these crop and grassland areas into homogeneous segments (spatial units). The segmentation method used is that of "mean-shift". This function is available in the Orfeo ToolBox (OTB) image processing library[6], developed by the French space agency. This segmentation function can be executed from the QGIS interface. The segmentation "mean-shift" produces a vector file composed of several features (of polygonal type) at the output. Each feature delimits a homogeneous spatial unit. The spatial unit may be sub-plot, a plot or a set of plots with close NDVI values.

The soil moisture value will be estimated mainly by using the mean of radar pixels delimited by the entity. A 10 m buffer inside each polygon is considered to ensure that the pixels used to calculate the mean of backscattered signal do not contain border pixels (hedges, road, etc.). Before applying the buffer zone, a smoothing to slightly round the polygon vertices is necessary in order to avoid topological problems[7]. In addition, to obtain a reliable average backscattered signal, features that delimit less than 20 radar pixels have been eliminated. Indeed, the averaging backscattering coefficients of less than 20 pixels is not relevant due to the speckle present on radar images. To eliminate polygons delimiting less than 20 pixels, the number of radar pixels that delimit each entity (spatial unit) is first calculated using the "Zonal statistics" function available in QGIS. The number of pixels within each feature is automatically recorded in the attribute table of the segmentation file. Then, features with less than 20 pixels are deleted. Part 4 of the flowchart (Figure 1.2) shows the procedures followed to segment the crop and grasslands areas, and remove features with less than 20 pixels.

QGIS functionality to activate OTB:
- *Processing > Options > Providers > Orfeo Toolbox >...*

QGIS functionality to create a mask:
- *Raster > Raster Calculator >...*

QGIS functionality to segment the NDVI image:
- *Processing > Toolbox > Orfeo Toolbox > Segmentation>Segmentation (meanshift) >...*

QGIS functionality to smooth the polygons:
- *Processing > Toolbox > QGIS geoalgorithms > vector geometry tools > smooth geometry >...*

QGIS functionality to do a buffer distance:
- *Vector > Geoprocessing Tools > Fixed Distance Buffer >...*

---

6 https://www.orfeo-toolbox.org.
7 https://docs.qgis.org/2.6/fr/docs/gentle_gis_introduction/topology.html.

QGIS functionality to calculate the number of pixels per band:
- *Raster > Zonal statistics > Zonal statistics >...*

QGIS functionality to delete a feature according to attribute value:
- *Right click on the vector layer > Open Attribute Table > Select features using an expression > Delete selected features >...*

1.3.3. *Soil moisture mapping*

Soil moisture mapping consists of estimating moisture at the level of each feature (spatial unit) for a given radar image. This soil moisture is derived mainly by using the mean of the radar pixels delimited by the feature.

To map the soil moisture at a given radar date, a zonal statistic (mean) is first performed on each feature (using the segmentation file), for each of the three radar bands (VV, VH and local incidence) and for the NDVI image. The mean values for each feature are automatically recorded in the attribute table of the segmentation file. Then, the mean values of each feature are exported and saved in a text file. In this text file, each row represents a feature, and the columns are in the order: (1) identification of the useful feature (crops or grasslands), (2) mean of the backscattered radar signal in VV polarization, (3) mean of the backscattered radar signal in HV polarization, (4) mean of the local incidence angle and (5) averaged NDVI. Finally, the inversion algorithm of the radar signal can be applied using this text file to estimate soil moisture for each useful feature (feature that delimits more than 20 radar pixels).

To apply the estimation algorithm of the soil moisture (launched using Python), it is necessary to install the latest version of the free Python[8] software and the associated libraries[9] (Scipy, numpy + mkl and keras). The algorithm, once applied, automatically creates a file called "results" and places it in the same directory as the text file that contains the input information. In this result file, the first column is the identifier of each entity and the second column represents the estimated soil moisture.

QGIS functionality to calculate the mean of pixels per band:
- *Raster > Zonal statistics > Zonal statistics >...*

QGIS functionality to export the zonal statistics as a text file:
- *Right click on the vector layer > Save As >...*

---

8 https://www.python.org.
9 http://www.lfd.uci.edu/%7Egohlke/pythonlibs.

To produce the moisture map, a join according to the identifier between the segmentation layer composed of the used features and the file containing the estimated moisture values "results" is performed. Finally, for a better visualization of the estimated moisture values, we have coded the estimates according to classes with a range of 5 vol.% (0–5 vol.%, 5–10 vol.%, etc.). Part 5 of the flowchart (Figure 1.2) shows the procedures for the production of soil moisture maps.

QGIS functionality to import a text file:
- *Layer > Add Layer > Add Delimited Text Layer > Format >...*

QGIS functionality to join according to an identifier a vector layer and a text file:
- *Right click on the vector layer > Properties > Joins >...*

QGIS functionality to **create a map showing for each entity the estimated soil moisture:**
- *Right click on the vector layer > Properties > Style >...*

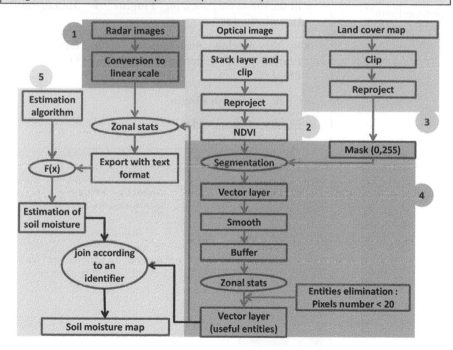

Figure 1.2. *Steps followed to produce the soil moisture maps. For a color version of the figure, see www.iste.co.uk/baghdadi/QGIS2.zip*

1.4. Implementation of the application via QGIS

This section contains the implementation of QGIS functions that lead to obtaining the soil moisture maps by coupling radar and optical data. For color versions of the figures in the following sections, see www.iste.co.uk/baghdadi/QGIS2.zip.

1.4.1. *Layout*

In this section, the steps to produce a map of the study site are presented.

Process	Practical implementation in QGIS
1. Visualization of the study site with a satellite image from Google Earth[10]	➤ Install a plugin to view a Google Earth background map in QGIS: • In the menu bar, click on "**Plugins**" ➔ "**Manage and Install Plugins**". • In the window that appears, type "**OpenLayers Plugin**" and select the extension "**OpenLayers Plugin**". • Click on "**Install Plugin**". • Finally, click on "**Close**".

---

10 https://www.google.fr/intl/fr/earth.

	➢ To display the background map in QGIS: • In the menu bar, click on **"Web"** ➔ **"OpenLayers plugin"** ➔ **"Google maps"** ➔ **"Google Satellite"**. ➢ To display a vector layer that delimits the study area: • In the menu bar, click on **"Layer"** ➔ **"Add a Layer"** ➔ **"Add Vector Layer"**. • In the window that appears, click on **"Browse"** and select the vector layer. This vector layer is available in the data provided with the chapter. • Zoom on the study area.
2. Creating a new print composer, and viewing the Google Earth map	➢ To create a new print composer: • In the main menu, click on **"Project"** ➔ **"New Print Composer"**, or type: **"Ctrl+P"**. • In the window that appears, give a name to the print composer, for example **"study area"**, and click on **"ok"**. ➢ Display the Google Earth image in the print composer: • In the main menu bar, click on **"Layout"** ➔ **"Add Map"**, and draw with the mouse a rectangular area in the existing sheet.
3. Insertion of a grid of geographical coordinates	➢ To set the size and the scale of the print composer: • In the menu bar, click on **"Layout"** ➔ **"Move Item"** and select the rectangular area containing the image. • In the section **"Composition"**, configure the sheet type to **"A4 (210 mm × 297 mm)"**. • In the section **"Item properties"**, go to **"Main properties"** and enter 100000 in **"Scale"** to define a scale of 1/100000.

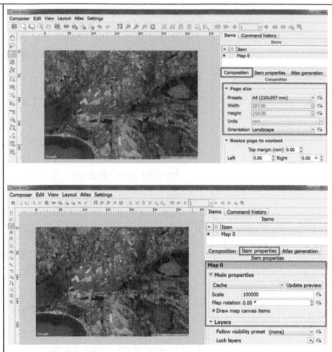

> ➤ To add a coordinate grid:

> • In the section "**Items properties**", go to "**Grids**" and
> click on "**+**" to add a coordinate grid. In addition,
> define the grid interval as 2 km along the *X*-axis and
> the *Y*-axis.

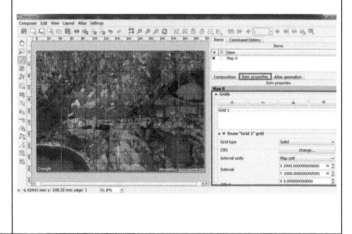

	➤ To display the Cartesian coordinates: • In the menu **"Item properties"**, go to **"Grids"**, and activate the option "**Draw coordinates**". • Disable the display of the coordinates at the top and at the right of the rectangle containing the image, and keep the display at the bottom and at the left. • Click on **"Font"**, and define font and size of coordinates. • On the field "**Coordinate precision**" enter 0. 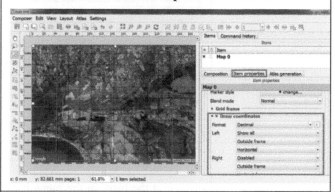
4. Insertion of the graphic scale, and the arrow of the North	➤ To add the graphic scale: • In the menu bar, click on **"Layout"** ➔ "**add scalebar**" • In "**Items properties**", go to "**Units**", "**Segments**", and "**Fonts and colors**" to define the unit, the number of segments and the font of the graphic scale, respectively. 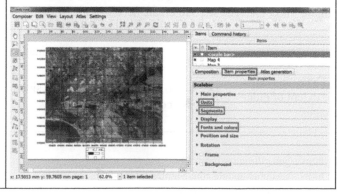

	➤ To add the North arrow:
	• Download from the Internet[11] an image of the North arrow, and then save this image on the computer.
	• In the menu bar, click on **"Layout"** ➜ **"Add Image"** and draw a rectangular area.
	• Select the rectangular area drawn. Then, in the field **"Item properties"**, go to **"Main properties"** and enter in **"Image source"** the image of the North arrow downloaded from the website.
	➤To print the map, in the menu bar, click on **"Composer"** ➜ **"Export as PDF"**.

Table 1.1. *Layout of the study site*

1.4.2. *Radar images*

1.4.2.1. *Download images*

To download the radar images, create first a Gmail[12] account and Google Earth Engine[13] account. The steps to create accounts are available on the webpages.

Process	Practical implementation in QGIS
1. Determination of the extent of radar images	➤ Go to Google Earth Engine site and click on **"platform"** ➜ **"code editor"**.

---

11 https://goo.gl/images/mns0lM.
12 https://accounts.google.com/SignUp?hl=en-GB.
13 https://earthengine.google.com.

> Click on **"New file"** to create a new script (called QGIS in the following screenshot).

> Draw a polygon that covers the study site with the tool **"Draw a shape"** located to the left of the following screenshot. This polygon represents the extent of the radar image to download.

2. Select and download radar images	> Search for available images: • Copy the following code in the text editor of the script (QGIS) to display all images acquired between 19/01/2017 and 27/01/2017. Please note, the following three code lines (marked in bold) should be on the same line in the text editor of the script. *var start = new Date("01/19/2017");* *var end = new Date("01/27/2017");* *var s1 = ee.ImageCollection('COPERNICUS/S1_GRD')*

```
        .filterDate(start, end)
        .filterBounds(geometry);
    var count = s1.size().getInfo();
    for (var i = 0; i < count ; i++) {
      var img = ee.Image(s1.toList(1, i).get(0));
      var geom = img.geometry().getInfo();
      Export.image(img, img.get('system:index').getInfo(),
      {'scale': 20,'crs': 'EPSG:4326','region':
      geometry.toGeoJSONString()});
    }
```

- Click on "**Run**", the images will appear on the right of the window, in the menu "**Tasks**". In this menu, we observe the two images that we seek to download.

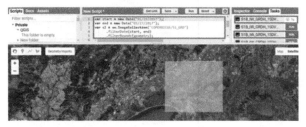

➢ To order the images:

- Click on "**Run**" in the field "**Tasks**".

- In the window that appears, enter the spatial resolution of images (10 m).

- Click on "**Run**". The images are transferred automatically to Google Drive[14], and it becomes possible to download them: S1B_IW_GRDH_1SDV_ 20170119T055130_20170119T055155_003913_ 006BD9_1D91.tif, and S1A_IW_GRDH_1SDV_ 20170126T173853_20170126T173918_015006_018823 _7F07.tif

Table 1.2. *Radar images downloading*

---

14 https://drive.google.com/drive/my-drive.

1.4.2.2. *Radar signal conversion to linear scale*

In order to apply the soil moisture estimation algorithm, it is important to convert the first two bands of the downloaded radar images of the dB scale into linear scale:

$$Band\_linear\_scale = 10^{\left(\frac{Band\_dB\_scale}{10}\right)} \qquad\qquad [1.4]$$

Process	Practical implementation in QGIS
1. Conversion into linear scale of the first two bands of the radar image acquired on 19/01/2017	First, the first two bands were converted one by one into a linear scale (there are in dB scale). Then, the two bands in linear scale and the incidence angle band were stacked to facilitate their use thereafter. ➢ Import the radar images into QGIS: • In the menu bar, click on **"Layer"** ➔ **"Add Layer"** ➔ **"Add Raster Layer"**. • In the window that appears, select the two radar images and click on **"ok"** to display the images in QGIS. ➢ To convert the first two bands of the radar image acquired on 19/01/2017 in a linear scale: • In the menu bar, click on **"Raster"** ➔ **"Raster Calculator"**. In the window that appears, the part **"Raster bands"** shows the three bands of the radar image acquired on 19/01/2017. • Type in **"Raster calculator expression"** the following equation to convert the band 1 of the radar image to a linear scale. Then, name the output image as follows: "20170119_b1_lin.tif". 10^("S1B_IW_GRDH_1SDV_20170119T055130_201701 19T055155_003913_006BD9_1D91@1"/10) • Click on **"ok"** to start the calculation.

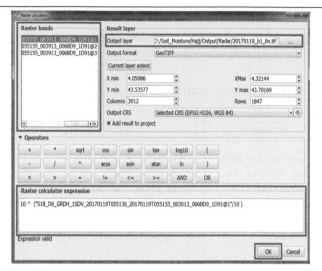

- To convert the second band of the radar image (19/01/2017) to a linear scale, repeat the above step. Name the output image as follows: "20170119_b2_lin.tif".

➤ To extract the band that corresponds to local incidence angle, type in "**Raster calculator expression**" the following equation. Name the output image as follows: "20170119_b3.tif"

"S1B_IW_GRDH_1SDV_20170119T055130_20170119T 055155_003913_006BD9_1D91@3"*1

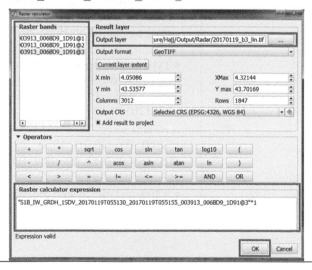

	➢ To stack the three single-band images ("20170119_b1_lin.tif", "20170119_b2_lin.tif", "20170119_b3.tif"):
	
	• Import into QGIS the three bands of the radar image: "20170119_b1_lin.tif", "20170119_b2_lin.tif" and "20170119_b3.tif"
	
	• In the part "**Layer**" of the QGIS interface, check that the order of the single-band images is as below. This is important to have in the output image, band 1 = "20170119_b1_lin.tif", band 2 = "20170119_b2 _lin.tif" and band 3 = "20170119_b3.tif"
	
	    o "20170119_b1_lin.tif"
	
	    o "20170119_b2_lin.tif"
	
	    o "20170119_b3.tif"
	
	• In the menu bar, click on "**Raster**" ➔ "**Miscellaneous**" ➔ "**Build Virtual Raster**" for stacking images.
	
	• In the window that appears, click the options "**Use visible raster layers for input**", and "**Separate**". Then, name the output file as follows: "20170119_lin.vrt".

2. Conversion of the first two bands of the second radar image acquired on 26/01/2017 in a linear scale	• To verify the order of the bands in the output image (20170119_lin.vrt), look at the text area of the function "**Build Virtual Raster**" (screenshot below). This text zone shows that in the output image, the bands 1, 2 and 3 are the single-band images "20170119_b1_lin.tif", "20170119_b2_lin.tif" and "20170119_b3.tif", respectively. • Finally, click on "**ok**". Once processing is complete, the output image "20170119_lin.vrt" appears in the "**Layers**" and "**Display**" part of QGIS.

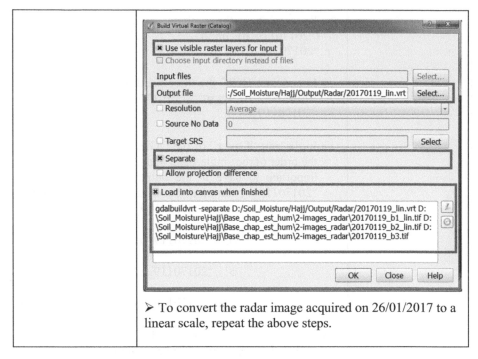

> To convert the radar image acquired on 26/01/2017 to a linear scale, repeat the above steps.

Table 1.3. *Radar images pre-processing*

1.4.3. *Optical image*

1.4.3.1. *Stacking, clipping and reprojection*

The optical image is downloadable as 13 separate spectral bands. To facilitate the use of the optical image, only the blue (band 2), green (band 3), red (band 4) and near-infrared (band 8) bands were stacked. Then, the image composed of the four bands is clipped. Finally, the clipped image is reprojected to have the same projection system as the radar images (WGS 84).

Process	Practical implementation in QGIS
1. Stacking of the four bands	➤ Download the optical image: • Download from the Theia[15] website the S2A image acquired on 15/10/2016 over Montpellier (south of France). The identifier of the optical image on the Theia Website is: SENTINEL2A_20161015-104513-300_L2A_T31TEJ_D • To access the 13 spectral bands of the optical image, unzip the downloaded image. ➤ Stack the four bands: • Import the four spectral bands into QGIS: o Blue: SENTINEL2A_20161015-104513-300_L2A_ T31TEJ_D_V1-1_FRE_B2.tif o Green: SENTINEL2A_20161015-104513-300_ L2A_T31TEJ_D_V1-1_FRE_B3.tif o Red: SENTINEL2A_20161015-104513-300_L2A_ T31TEJ_D_V1-1_FRE_B4.tif o Near-infrared: SENTINEL2A_20161015-104513-300_L2A_T31TEJ_D_V1-1_FRE_B8.tif

---

15 https://theia.cnes.fr/atdistrib/rocket/#/search?collection=SENTINEL2.

	• To stack the four bands, follow the steps detailed in section 1.4.2.2. Name the output image as follows: "20151510.vrt". In the output image the bands 1, 2, 3 and 4 are, respectively, the Blue (band 2), the Green (band 3), the Red (band 4) and the Near-Infrared (band 8) of the optical S2A image. 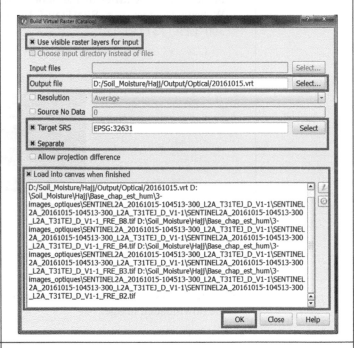
2. Clip the stacked image	➤ To clip the stacked image according to the extent of the study site: • Import into QGIS the image 20151510.vrt and the vector layer delimiting the study area. • In the menu bar, click on **"Raster"** ➔ **"Extraction"** ➔ **"Clipper"**. • In the window that appears, select in **"Input file"** the stacked image (20151510.vrt). • In **"Output file"**, give to the output image the following name: 20161015_clip.tif. • Click the option **"Mask Layer"**, and select the vector layer "**zone_etude.shp**".

	• Activate the option "**Clip the extent of the target dataset to the extent of the cutline**". • Finally, click on "**ok**". Once the processing is complete, the clipped image appears in the "**Layers**" and "**Display**" part of the QGIS interface. 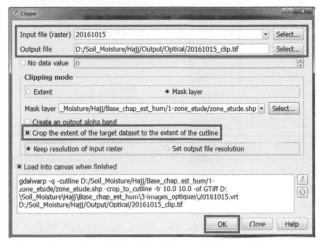
3. Reprojection of the stacked and clipped image	➢ To reproject the image: • In the menu bar, click on "**Raster**" ➔ "**Projections**" ➔ "**Warp (Reproject)**". • In the window that appears, select in "**Input file**" the image 20161015_clip.tif. • Activate the option "**Target SRS**", and choose the projection WGS 84 (EPSG: 4326). • In "**Output file**", give to the output image the following name: 20161015_clip_rep.tif. • Finally, click on "**ok**". Once the processing is complete, the output image appears in the "**Layer**" and "**Display**" part of the QGIS interface.

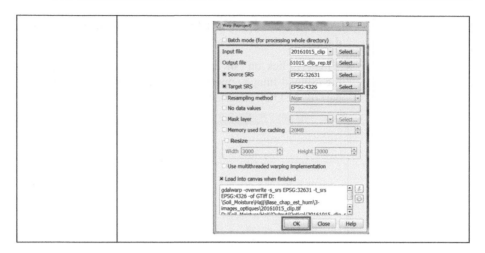

Table 1.4. *Optical image pre-processing*

1.4.3.2. *Calculation of the NDVI image*

The NDVI image is derived from the optical image (20161015_clip_rep.tif). This NDVI image is segmented to delimit the spatial units (polygons).

Process	Practical implementation in QGIS
1. Calculation of the NDVI image	➢ To calculate the NDVI: • Import in QGIS the image 20161015_clip_rep.tif. • In the menu bar, click on "**Raster**" ➔ "**Raster Calculator**". • In the window that appears, type the following formula. With this formula, the NDVI values are multiplied by 100. This allows the encoding of the NDVI image to be converted into 16 bits without losing precision, and consequently accelerates the segmentation of the NDVI image. 100*(("20161015_clip_rep@4" − "20161015_clip_rep@3") / ("20161015_clip_rep@4" + "20161015_clip_rep@3")) • In "**Output Layer**", name the image as: "NDVI_20161015.tif". • Finally, click on "**ok**". Once processing is complete, the

output image appears in the "**Layer**" and "**Display**" part of the QGIS interface.

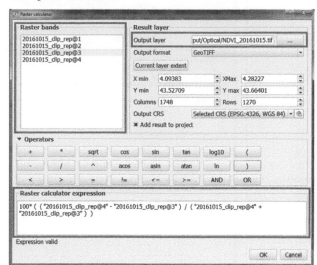

➤ To convert the NDVI image (NDVI_20161015.tif) into a 16-bit integer:

- In the menu bar, click on "**Raster**" ➔ "**Conversion**"➔ "**Translate (Convert Format)**".

- In the window that appears, select in "**Input Layer**" the NDVI image (NDVI_20161015.tif).

- In "**Output file**" enter NDVI_20161015_ent16.tif.

- Activate the option "No data" and enter –32768.

- Click on "**Edition**" and add "*-ot Int16*" in the text area immediately after "*gdal_translate -a_nodata -32768 -of GTiff*".

- Finally, click on "**ok**". Once processing is complete, the output image appears in the "**Layer**" and "**Display**" part of the QGIS interface.

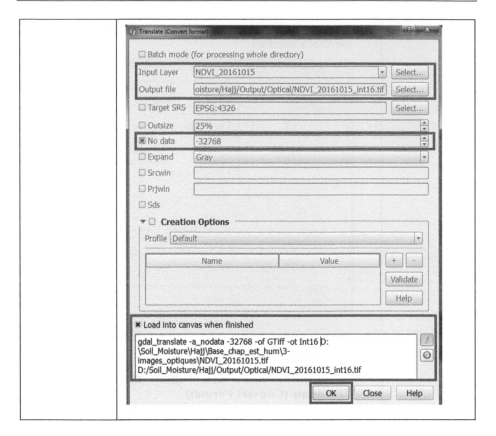

Table 1.5. *Calculation of the NDVI*

1.4.4. *Land cover map*

To facilitate the manipulation of the land cover map, this map is first clipped, and then reprojected to WGS 84. Procedures for clipping and reprojecting an image are explained earlier in section 1.4.3. The clipped and reprojected land cover map will be called ocsol_clip_wgs84.tif.

1.4.5. *Segmentation of crop's areas and grasslands*

To delimit the spatial units, a mask is first generated from the land cover map to determine the crop and grasslands areas. Then, the NDVI image is used to segment crop and grassland areas into homogeneous segments (spatial units).

Process	Practical implementation in QGIS
1. Determination of crop's areas and grasslands, and segmentation	➢ To create the mask image: • Import the clipped and reprojected land cover map into QGIS (ocsol_clip_wgs84.tif), as well as the NDVI image (NDVI_20161015_ent16.tif). The import of the NDVI image allows obtaining the mask image with an extent and a spatial resolution identical to those of the NDVI image: NDVI_20161015_ent16.tif. To do so, in "**Raster bands**" select the image NDVI_20161015_ent16.tif and then click on "**Current layer extent**". • In the menu bar, click on "**Raster**" ➔ "**Raster Calculator**". • In the window that appears, type the following formula and name the output image: mask.tif. ("ocsol_clip_wgs84@1" = 11 OR "ocsol_clip_wgs84@1" = 12 OR "ocsol_clip_wgs84@1" = 211)*255 Using this formula, the pixels of the land cover map (ocsol_clip_wgs84.tif) with values equal to 11 (summer crop), 12 (winter crop) and 211 (permanent grasslands) are set to 255 in the mask image (mask.tif), while the other pixels are set to 0.

➤ To segment the areas of the NDVI image that correspond to crop plots and grasslands:

- Import the mask image (mask.tif) and the NDVI image (20161015_ent16.tif).

- In the menu bar, click on **"Processing"** ➔ **"Toolbox"**.

- In the window that appears to the right of the QGIS interface, click on **"Orfeo Toolbox (Image Analysis)"** ➔ **"Segmentation"** ➔ **"Segmentation (meanshift)"**.

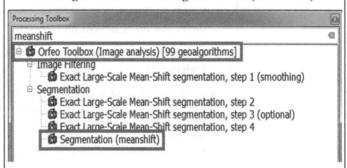

- In the window that appears, select the input image (**"input image"**: "NDVI_20161015_ent16.tif"), and define the spatial radius (**"spatial radius"** = 30 pixels), the range radius (**"range radius"**: 10) and the mask image (**"mask image"**: "masque.tif").

- Name the vector layer of the output segmentation as follows: **"output vector file"**: "seg_crops_grass.shp".

- Click on **"run"** to execute the segmentation function. Once processing is complete, the vector layer of the segmentation appears in the **"Layers"** and **"Display"** part of the QGIS interface.

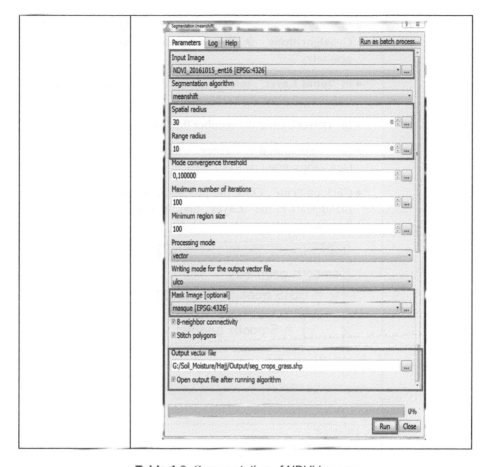

Table 1.6. *Segmentation of NDVI images*

1.4.6. *Elimination of small spatial units*

To eliminate the small spatial units present in the segmented image, a smoothing of the polygon is first applied. A buffer zone of 10 m (0.0001°) inside each polygon is produced. Then, the number of pixels in each polygon is calculated. Finally, polygons with less than 20 pixels are deleted.

Process	Practical implementation in QGIS
1. Elimination of small spatial units	➢ To smooth polygons: • In the main menu, click on "**Processing**" ➔ "**Toolbox**".

- In the window that appears, click on "**QGIS Geoalgorithms**" ➔ "**Vector geometry tools**" ➔ "**Smooth geometry**".

- In "**Input Layer**", select the vector layer of the segmentation (seg_crops_grass.shp).

- In "**Offset**", enter the number 0.5 (unit = pixel).

- In "**Smoothed**", enter the output file name (seg_crops_grass_smooth.shp).

- Click on "**run**". Once the processing is complete, the output file appears in the "**Layer**" and "**Display**" part of the QGIS interface.

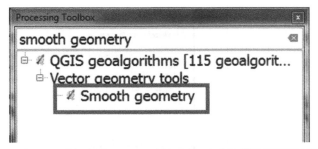

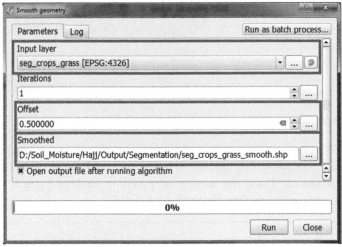

➤ To perform a buffer zone of 10 m:

- In the menu bar, click on "**Vector**" ➔ "**Geoprocessing Tools**" ➔ "**Fixed distance buffer**".

- In the window that appears, select the input layer (seg_crops_grass_smooth.shp).

- In "**Distance**", enter the number –0.0001° (–10 m).

- In "**Buffer**", enter the output file name (seg_crops_grass_smooth_buff-10.shp).

- Click on "**run**". Once the processing is complete, the output file appears in the "**Layers**" and "**Display**" part of the QGIS interface.

Please note that the "buffer" function automatically eliminates all polygons with width smaller than 20 m.

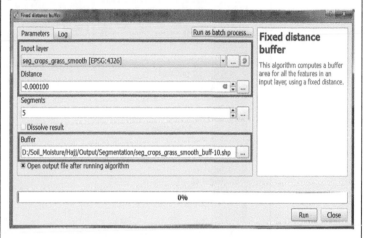

➢ To calculate the number of pixels in each polygon:

- Install the plugin "**Zonal statistics**". The installation procedures for a plugin are already explained in section 1.4.1.

- Import in QGIS the radar image 20170119_lin.vrt, and the vector layer seg_crops_grass_smooth_buff-10.shp.

- In the menu bar, click on "**Raster**" ➔ "**Zonal Statistics**" ➔ "**Zonal Statistics**".

- In "**Raster layer**", select the radar image 20170119_lin.vrt.

- In "**Band**", select the band 1 (VV polarization in linear scale).

- In "**Polygon layer containing the zones**", select the vector layer seg_cultures_prairie_liss_buff-10.shp.

- In "**Output column prefix**", type "*nb*".

- In "**Statistics to Calculate**", check only the option "**Count**".

- Click on "**ok**". Once the processing is complete, statistics (number of pixels) are saved in the attribute table.

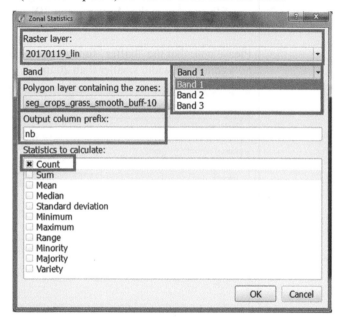

➢ To remove entities with a number of pixels less than 20:

- Open the attribute table of the vector layer seg_cultures_prairie_liss_buff-10.shp: "**right click**" on the vector layer ➔ "**Open attribute table**".

- Activate the edit mode 🖉 .

- Click on "**Select features using an expression**" 🖾 and type "**"nbcount"** < *20*".

- Click on the icon "delete selected features" 🗑 .

- Disable the edit mode.

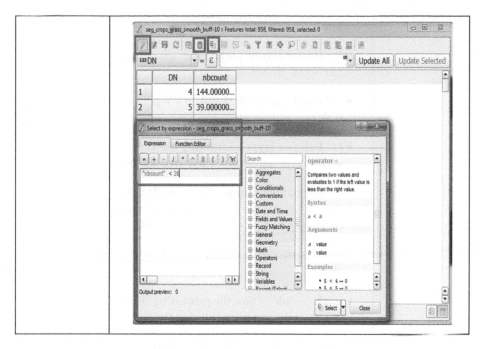

Table 1.7. *Emimination of small spatial units*

1.4.7. *Mapping soil moisture*

1.4.7.1. *Calculation of mean backscattered signal, mean incidence angle and mean NDVI*

For each spatial unit (polygon), the mean of the backscattered radar signal (linear scale) in VV and HV, of the local incidence angle, and of the NDVI, was calculated. Then, the mean values are exported to csv.

Process	Practical implementation in QGIS
1. Calculation of mean backscattered radar signal, mean incidence angle and mean NDVI for each spatial unit	➢ Calculate the mean backscattered radar signal (linear scale) using all pixels containing in each polygon (spatial unit): • Import the radar image acquired on 19/01/2017 (20170119_lin.vrt), and the segmentation vector layer segmentation seg_crops_grass_smooth_buff-10.shp. • In the main menu, click on **"Raster"** ➔ **"Zonal**

Statistics" ➔ "**Zonal Statistics**".

- In "**Raster Layer**", select the radar image 20170119_lin.vrt.

- In "**Band**", select the band 1 (VV polarization in linear scale).

- In "**Polygon layer containing the zones**", select the vector layer seg_crops_grass_smooth_buff-10.shp.

- In "**Output column prefix**", type "MVV_170119" to denote that it is the mean (M) of the radar backscattering coefficient in VV for the of 19/01/2017.

- In "**Statistics to calculate**", check only the option "Mean".

- Click on "**ok**". Once the processing is complete, the values of the mean are saved in the attribute table.

➢ Repeat the above step for band 2 (HV in linear scale: MHV_170119) and the band 3 (incidence angle: MI_170119) of the image 20170119_lin.vrt.

➢ Repeat these steps for the three bands of the image 20170126_lin.vrt.

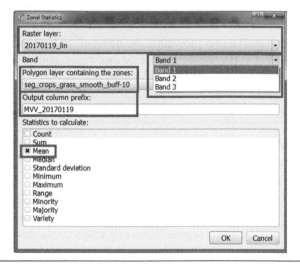

> ➤ Open the attribute table to check the results.

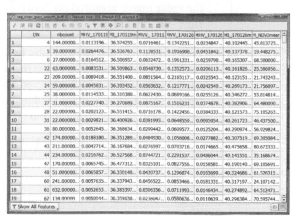

> ➤ Calculate the mean of the NDVI using pixels containing in each polygon (spatial unit):

- Import the NDVI image (NDVI_20161015_ent16.tif) and the vector layer seg_crops_grass_smooth_buff-10.shp.

- Calculate the mean NDVI for each entity (spatial unit) according the steps detailed above. Name the field "M_NDVImean".

> ➤ To export the values from the vector layer to .csv file:

- Import the segmentation vector layer seg_cultures_prairie_liss_buff-10.shp.

- Right click on the vector layer seg_cultures_prairie_liss_buff-10.shp, and select "**Save As…**".

- In "**Format**", select the option "**Comma Separated Values csv**".

- In "**File name**", enter the following name stat_20170119.csv.

- Check the following fields: DN, MVV_170119, MHV_170119, MI_170119m, M_NDVI_mean.

- Click on "**ok**".

> ➤ Repeat the above step to export mean values calculated on the radar image acquired on 26/01/2017 into another csv file (stat_170126.csv).

> ➤ Please note, in each file .csv, the first, second, third and fourth columns should be in the order: the identifier of each segment (DN), backscattered radar signal in VV (linear scale), backscattered radar signal in HV (linear scale), local incidence angle and NDVI.

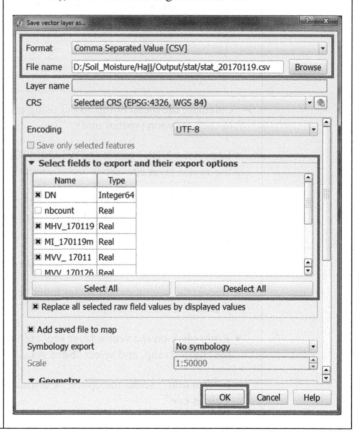

Table 1.8. *Calculation of mean backscattered signal, mean incidence angle and mean NDVI*

1.4.7.2. *Execution of the moisture estimation algorithm*

1.4.7.2.1. Installing Python and associated libraries

Process	Practical implementation in QGIS
1. Installing Python	➤ Download the Python[16] executable, and run the installation. In **"Customize installation"**, keep the options by default, and change the installation directory to "C:\Python36".
2. Installing the libraries	➤ In **"Environment Variables"** of the computer, add the path to "python.exe": • Click on **"Start"**. • In **"Search Program and file"**, click on the link: **"View advanced system settings"**. • Click on **"Environment Variables"**. In the section **"User Variables"**, select the variable environment **"PATH"**. • Click on **"Edit"**, and enter the path of "python.exe": "C:\Python36", and of "pip.exe": "C:\Python36\Scripts".

16 https://www.python.org.

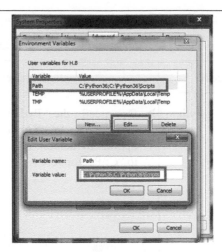

> In "**Search Program and files**", open the window of "**Command Prompt**", and type "python" to check if python is now an executable program.

> Download the libraries "Scipy et numpy+mkl"[17], and put them in "C:\Python36\Scripts".

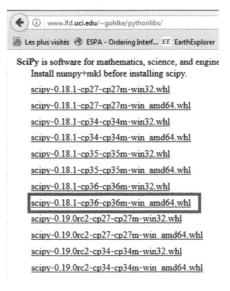

_____

17 http://www.lfd.uci.edu/%7Egohlke/pythonlibs.

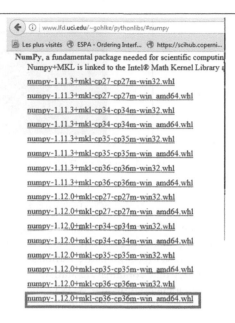

➢ To install the libraries, open a window of "**Command Prompt**" and:

• Type "pip install C:\Python36\Scripts\scipy-0.18.1-cp36-cp36m-win_amd64.whl" to install the "Scipy" module.

• Type "pip install keras" to install the "keras" module.

• Type "pip install C:\Python36\Scripts\numpy-1.12.0+mkl-cp36-cp36m-win_amd64.whl" to install the numpy+mkl module.

Table 1.9. *Installing Python and associated libraries*

1.4.7.2.2. Application of the soil moisture estimation algorithm

To apply the soil moisture estimation algorithm, you should use the script consisting of programming codes written in Python language (apply_algo.py) to run the algorithm algo.sav (soil moisture estimation by neural network). The script uses a .csv file as input that contains the statistics (mean of radar signal in VV and HV polarizations, incidence angle and NDVI) and produces a text file that contains the identifier of each spatial unit and the associated estimated moisture.

Process	Practical implementation in QGIS
1. Execution of the algorithm to estimate the moisture on each polygon	➢ To apply the soil moisture estimation algorithm: • Move the csv files to the same directory as the script (apply_algo.py). • Open the script (apply_algo.py): "**Right Click on the script**" ➔ "**Edit with IDLE**". • In the "zonal_stat" variable, put the name of the file .csv containing the statistics: zonal_stat = "stat_20160119.csv". • Click on "**F5**" on the keyboard to launch the algorithm. • A text file is automatically generated (results_stat_20160119.txt). In this text file, the first column is the identifier (DN), and the second column is the estimated soil moisture. ➢ Repeat the steps below to estimate the soil moisture using the file csv: "stat_20170126.csv". The results are saved automatically in a file called: "results_stat_20160126.txt".

Table 1.10. *Estimation of soil moisture*

1.4.7.3. *Production of soil moisture maps*

To produce soil moisture maps, joins are made according to the identifier (DN) between the text files that contain the estimated moisture values and the vector layer seg_crops_grass_smooth_buff-10.shp. This allows the estimated moisture values for the dates of the radar images (19/01/2017 and 26/01/2017) to be added in the attribute table of the vector layer seg_crops_grass_smooth_buff-10.shp. Then, visualization in the form of a map of the estimated moisture values is carried out.

Process	Practical implementation in QGIS
1. Join according to the identifier and viewing maps	➢ Add the estimated moisture values in the attribute table of the vector layer seg_crops_grass_smooth_buff-10.shp. • Import the vector layer seg_crops_grass_smooth_buf-10.shp and the text file results_stat_20160119.txt. To import the last file, click on "**Add Layer**" ➔ "**Add Delimited Text Layer**" in the menu bar. 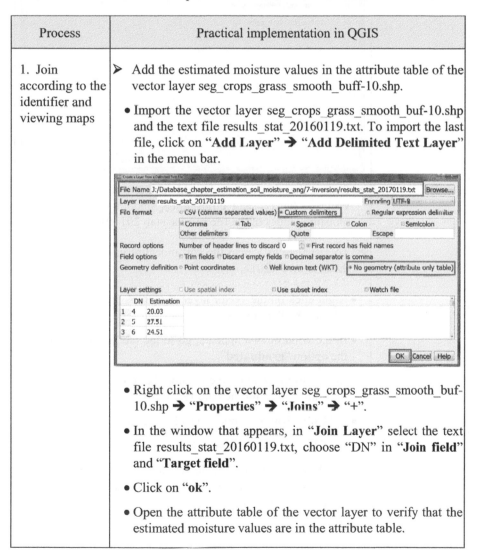 • Right click on the vector layer seg_crops_grass_smooth_buf-10.shp ➔ "**Properties**" ➔ "**Joins**" ➔ "**+**". • In the window that appears, in "**Join Layer**" select the text file results_stat_20160119.txt, choose "DN" in "**Join field**" and "**Target field**". • Click on "**ok**". • Open the attribute table of the vector layer to verify that the estimated moisture values are in the attribute table.

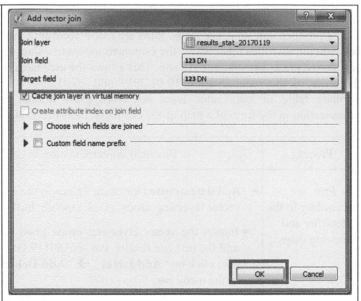

➤ Repeat the above steps to add in the attribute table of the vector layer seg_crops_grass_smooth_buf-10.shp the moisture values of the file results_stat_20170126.txt.

➤ To visualize as a map the soil moisture values estimated for the date 19/01/2017:

• Right click on the vector layer seg_crops_grass_smooth_buf-10.shp ➔ "**Properties**".

• In the window that appears, go to the tab "**style**" and choose the option "**graduated**".

• In "**Column**", choose the column that contains the estimated moisture values for the date 19/01/2017.

• In "**Color Ramp**", choose a Color degradation.

• In "**Classes**", define seven classes.

• In "**Values**", define the lower and upper bounds of each class. In the following figure, the classes are [5-10], [10-15], …, [35-40] (vol.%).

• Click on "**Ok**".

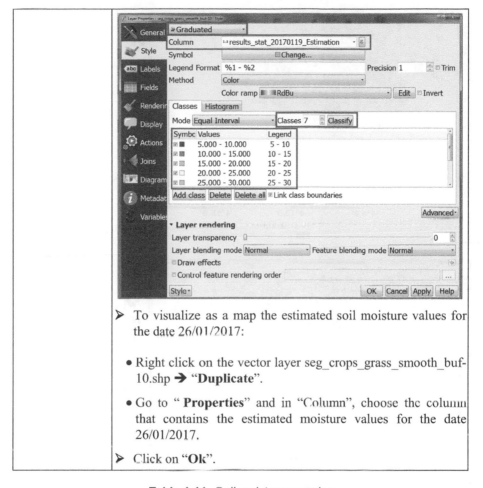

> To visualize as a map the estimated soil moisture values for the date 26/01/2017:

- • Right click on the vector layer seg_crops_grass_smooth_buf-10.shp → **"Duplicate"**.

- • Go to " **Properties**" and in "Column", choose the column that contains the estimated moisture values for the date 26/01/2017.

> Click on "**Ok**".

Table 1.11. *Soil moisture mapping*

1.4.8. *Soil moisture maps*

This section presents a spatial visualization of the estimated soil moisture values at each spatial unit (polygon). Figure 1.3 shows the soil moisture maps for the two radar images 19/01/2017 and 26/01/2017. The map of 19/01/2017 shows that the soil of the study site is dry with moisture values between 5 and 15 vol.%. The map of 26/01/2017 shows wet soils with moisture values between 25 and 35 vol.%.

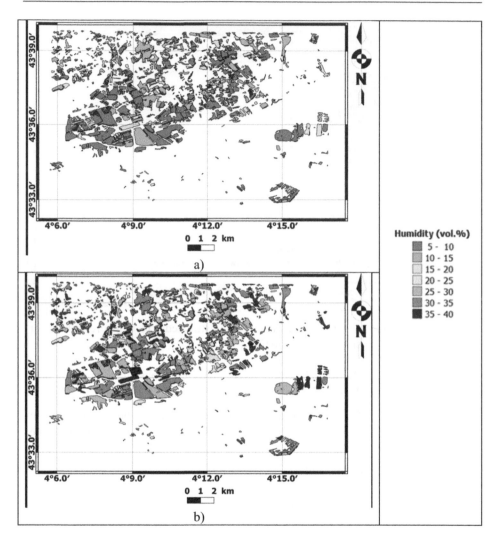

Figure 1.3. *Soil moisture maps for (a) 19/01/2017 and (b) 26/01/2017*

1.5. Bibliography

[ATT 78] ATTEMA E.P.W., ULABY F.T., "Vegetation modeled as a water cloud", *Radio Science*, vol. 13, pp. 357–364, 1978.

[BAG 06] BAGHDADI N., HOLAH N., ZRIBI M., "Calibration of the integral equation model for SAR data in C-band and HH and VV polarizations", *International Journal of Remote Sensing*, vol. 27, pp. 805–816, 2006.

[BAG 16a] BAGHDADI N., CHOKER M., ZRIBI M. *et al.*, "A new empirical model for radar scattering from bare soil surfaces", *Remote Sensing*, vol. 8, p. 920, 2016.

[BAG 16b] BAGHDADI N., ZRIBI M., "Characterization of soil surface properties using radar remote sensing", in BAGHDADI N., ZRIBI M. (eds), *Land Surface Remote Sensing in Continental Hydrology*, ISTE Press, London and Elsevier, Oxford, 2016.

[DUB 95] DUBOIS P.C., VAN ZYL J., ENGMAN T., "Measuring soil moisture with imaging radars", *IEEE Transactions on Geoscience and Remote Sensing*, vol. 33, pp. 915–926, 1995.

[EL 09] EL HAJJ M., BÉGUÉ A., GUILLAUME S. *et al.*, "Integrating SPOT-5 time series, crop growth modeling and expert knowledge for monitoring agricultural practices – the case of sugarcane harvest on Reunion Island", *Remote Sensing of Environment*, vol. 113, pp. 2052–2061, 2009.

[EL 16] EL HAJJ M., BAGHDADI N., ZRIBI M. *et al.*, "Soil moisture retrieval over irrigated grassland using X-band SAR data", *Remote Sensing of Environment*, vol. 176, pp. 202–218, 2016.

[FUN 94] FUNG A.K., *Microwave Scattering and Emission Models and Their Applications*, Artech House, 1994.

[MIR 15] MIRANDA N., *Sentinel-1A TOPS Radiometric Calibration Refinement*, European Space Agency, Paris, 2015.

Disaggregation of Thermal Images

2.1. Definition and context

Satellite images are characterized by their spatial, temporal and spectral resolutions. Spectral resolution is linked to spectral bandwidth and the number of bands; narrow spectral bands can detect some characteristics of the surface that are not detectable by wider bands. Spatial resolution refers to the pixel size that ranges from a few centimeters to several kilometers. Temporal resolution is the overpass frequency of a satellite over the same point in the Earth. The overpass frequency of satellites may vary between a few hours and several days. For example, the spatial resolution of Sentinel-2A (S2A) images is 10–60 m depending on the spectral band, and the temporal resolution is 10 days over France (this can be reduced to 5 days by combining S2A and Sentinel-2B (S2B) since they have the same characteristics). Frequently, satellites offering a high spatial resolution have a low temporal resolution. However, for certain studies, images with high spatial and temporal resolution are necessary. Fusion methods have been developed to simulate time series with high spatial and temporal resolution by combining images from two different satellites: one offering low spatial resolution images but high temporal resolution and the other one having high spatial resolution but low temporal resolution [BIS 15, GAO 06].

For thermal images, the problem is quite different since thermal images usually have lower spatial resolution than optical images in the visible and near-infrared (VNIR). Moreover, not all satellites have bands in the thermal infrared (TIR) domain. For example, MODIS sensor offers daily images with a spatial resolution of

---

Chapter written by Mar BISQUERT and Juan Manuel SÁNCHEZ.

250–500 m in the optical domain but only 1,000 m in the thermal domain. Disaggregation approaches for thermal images have been developed to enhance their spatial resolution. These approaches allow the user, for example, to obtain disaggregated thermal images with 250 m spatial resolution when combining the information from the thermal and optical images provided by MODIS. The use of disaggregation methods applied to images from different satellites is also possible [BIS 16a, BIS 16b]. For example, the information in the optical domain from Sentinel-2 (S2) images with a spatial resolution of 10–60 m could be used in the disaggregation of thermal images from other sensors (MODIS, Sentinel-3, etc.).

2.2. Disaggregation method

In this chapter, a disaggregation method based on the linear relationship between optical and thermal information of the surfaces is applied. More specifically, the relationship between the vegetation index NDVI (Normalized Difference Vegetation Index) and the land surface temperature (LST) is used. This method had been originally developed to be applied to images from the same satellite [AGA 07]. More complex disaggregation methods based on data mining [GAO 12] or neural networks [BIN 13] have also been developed. However, it has been shown that for the disaggregation of MODIS thermal images (1 km) to Landsat spatial resolution (60 m), the method based on the linear relationship between NDVI and LST leads to the best results [BIS 16a].

Disaggregation methods aim to obtain thermal images with high spatial resolution (HR) from the combination of a thermal image with low spatial resolution (LR) and a VNIR image at HR. The method shown in this chapter also needs VNIR LR images from the same sensor as the LR image. The first step in the disaggregation method is to obtain the linear regression between LST and NDVI (equation [2.1]) from the LR images, for example from MODIS:

$$LST_{LR} = a + b\,NDVI_{BR} \qquad [2.1]$$

Under the assumption that the relationship between NDVI and LST is the same at LR and at HR, a and b parameters obtained from the LR images are applicable to the NDVI HR (e.g. S2) to simulate an HR thermal image.

2.2.1. Image pre-processing

To apply the disaggregation method, it is necessary to obtain three images on the same day: LR NDVI, LR LST and HR NDVI.

2.2.1.1. *Reproject, obtain the NDVI, mask and subset*

When using images from different satellites, it is possible that the cartographic projection is not the same. In this case, the images of one of the two satellites must be geo-referenced in the same way as the other satellite.

Since the disaggregation method is based on the relationship between NDVI and LST, the NDVI for both satellites ($NDVI_{LR}$ and $NDVI_{LR}$) must be calculated:

$$NDVI = \frac{\rho_{NIR} - \rho_{Red}}{\rho_{NIR} + \rho_{Red}} \hspace{3cm} [2.2]$$

where ρ_{NIR} is the reflectivity in the near-infrared band and ρ_{Red} is the reflectivity in the red band. In the image acquisition process from satellites, there may be some perturbing elements, such as clouds, shadows, etc. The presence of clouds pollutes the images since the recorded radiance comes from the clouds instead of from the surface of interest. Satellite products usually provide a quality mask with information about the presence of clouds and other problems. Therefore, the next step is to use this mask in order to eliminate polluted pixels.

On the other hand, thermal images from MODIS require the application of a scale factor ($\times$ 0.02) to have the temperature in degrees K (LST_{LR}).

Moreover, since the extension covered by images from different satellites is not the same, it is necessary to subset the images in order that all the images being used (in our case $NDVI_{HR}$, $NDVI_{LR}$ and LST_{LR}) cover the same extension.

The first part of the organigram (Figure 2.1) shows the pre-processing steps: reprojecting, obtaining the NDVI and LST, masking and subsetting.

QGIS function for obtaining the NDVI:
- Raster > Raster Calculator...

QGIS function for changing the projection:
- Raster > Projections > Warp (Reproject)...

QGIS function for subsetting all the images to the same extent:
- Raster > Align Rasters...

2.2.1.2. *NDVI normalization*

Since there may be some differences between the bands of the two sensors (due to the different wavelengths in each band and the spectral resolution), a normalization procedure can be applied to one of the two NDVI images. To apply the normalization, it is necessary to have, in the first instance, both images with the same spatial resolution. Therefore, the $NDVI_{HR}$ is first resized to the LR. The resizing is achieved by obtaining, for each LR pixel, the average NDVI of all the HR pixels inside the LR pixel: this image is named $NDVI_{HR}^{Av}$. Next, a linear regression is obtained between both images:

$$NDVI_{LR} = a1 + b1\, NDVI_{HR}^{Av} \qquad [2.3]$$

The normalized HR NDVI image ($NDVI_{HRnorm}$) is obtained by applying the following equation:

$$NDVI_{HRnorm} = a1 + b1\, NDVI_{HR} \qquad [2.4]$$

The second part of the organigram (Figure 2.1) shows this step of NDVI normalization.

QGIS function for resampling:
- *Processing > Toolbox > GRASS GIS Commands > Raster > r.resamp.stats* and *r.resamp.interp*

QGIS function for doing a linear regression:
- *Processing > Toolbox > GRASS GIS Commands > Raster > r.regression.line*

2.2.2. Disaggregation

2.2.2.1. First simulation of the HR LST

The disaggregation method is based on a pre-established relationship between LST and NDVI obtained from the LR images. Ideally, this relationship should be obtained from pure pixels covering homogeneous areas. To determine the homogeneity of each pixel, the coefficient of variation (CV, equation [2.5]) of each LR pixel can be obtained using the pixels from the $NDVI_{HR}$ image, which are inside each $NDVI_{LR}$ pixel. For each pixel "i" in the $NDVI_{LR}$ image, the CV is obtained as the standard deviation of the $NDVI_{HR}$ pixels divided by the average $NDVI_{HR}$ (pixels belonging to a pixel in the $NDVI_{LR}$ image):

$$CV_{LR\,i} = \frac{\sigma(NDVI_{HR})_i}{Average(NDVI_{HR})_i} \qquad [2.5]$$

To have a robust equation, pixels from the full range of NDVI values (0–1) should be used. These pixels are used for obtaining the relationship between LST_{LR} and $NDVI_{LR}$.

Once the equation between LST_{LR} and $NDVI_{LR}$ (equation [2.1]) is obtained from the homogeneous pixels in the LR image, this equation is applied to the $NDVI_{HRnorm}$ to obtain a first estimation of the simulated temperature at HR (LST'_{HR}):

$$LST'_{HR} = a + b \times NDVI_{HRnorm} \qquad [2.6]$$

The third part of the organigram (Figure 2.1) shows the steps followed to obtain the first simulation of LST at HR.

2.2.2.2. Correction of the simulated LST HR

Local effects (due, for example, to the climatic or microclimatic effects, to land use, soil humidity, etc.) may be present in the LST'_{HR} image. A correction can be applied to reduce these effects. First, the LST'_{HR} image is resized to the LR (LST^{Av}_{HR}) so it can be compared to the reference image (LST_{LR}). Next, an LR residual image is obtained as the difference between the resized simulated image (LST^{Av}_{HR}), and the reference LST image (LST_{LR}):

$$Residual = LST^{Av}_{HR} - LST_{LR} \qquad [2.7]$$

Next, this correction will be used to rectify the LST'_{HR} image. It is necessary to first resize the residual image, using, for example, the nearest neighbor approach (this means that all the HR pixels belonging to the same LR pixel will have the same value), to obtain the residual image at HR ($Residual_{HR}$). The HR residuals obtained from the LR image may cause a boxy effect that highlights the borders between neighboring LR pixels if there is an important radiometric difference between both pixels (see Figure 2.2). To minimize this boxy effect, a smoothing procedure can be applied to the image, for example a Gaussian filter ($Residual_{HRsm}$). The final HR temperature will be the first estimation of LST (LST'_{HR}) subtracting the smoothed residual image ($Residual_{HRsm}$):

$$LST_{HR} = LST'_{HR} - Residual_{HRsm} \qquad [2.8]$$

The fourth part of the organigram (Figure 2.1) shows the steps followed in the residual correction of the simulated image.

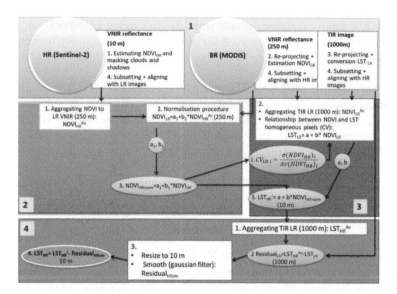

Figure 2.1. *Organigram of the processing chain for the disaggregation of thermal images. For a color version of the figure, see www.iste.co.uk/baghdadi/qgis2.zip*

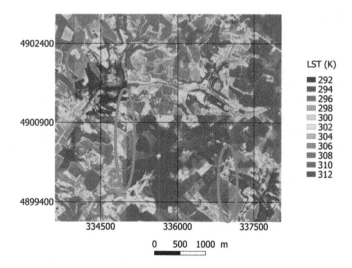

Figure 2.2. *Examples of the boxy effect. The green ellipses show the border effects between LR pixels (strong discontinuity between the radiometry of neighboring pixels). The coordinates correspond to the UTM WGS84 zone 31N projection. For a color version of the figure, see www.iste.co.uk/baghdadi/qgis2.zip*

2.3. Practical application of the disaggregation method

This section shows the practical application of the disaggregation of thermal images using MODIS and S2 images over part of the French territory, between the departments of Dordogne, Lot and Lot-et-Garonne. The images used include thermal images from MODIS (LST_{LR}) and reflectance images in the red and near-infrared domains from MODIS ($NDVI_{LR}$) and S2 ($NDVI_{HR}$).

The processing steps used correspond to the QGIS (version 2.18) software. All the steps could also be performed with better tools, such as using programming languages (Python, IDL, etc.).

2.3.1. Input data

This chapter is based on free data, downloadable from Internet. On one hand, reflectance and thermal MODIS images (products MOD09GQ and MOD11Q1, respectively), and on the other hand reflectance images from the S2 sensor that have been preprocessed by Theia (http://www.theia-land.fr). All the images correspond to the same date, September 28, 2016.

2.3.1.1. Downloading MODIS images

Reflectance and thermal images have been used here. The tile used for the MODIS images is "h18v04". Product MOD09GQ provides reflectance in the red and near-infrared domains needed to obtain the NDVI (at 250 m spatial resolution). The name of the file is "MOD09GQ.A2016272.h18v04.006.2016274063955.hdf". Product MOD11A1 provides the thermal images (to convert the image to K, a scale factor of 0.02 needs to be applied) at 1 km spatial resolution. The name of the file used is "MOD11A1.A2016272.h18v04.006.2016273090630.hdf".

Link: https://reverb.echo.nasa.gov/reverb

2.3.1.2. Downloading S2 images

S2 images have many spectral bands at different spatial resolutions. In this chapter, the red and near-infrared bands at 10 m spatial resolutions have been used. The name of the file to download is: "SENTINEL2A_20160928-105637-665_L2A_T31TCK_D_V1-0". Within the file, there are several files corresponding to the different bands and to the cloud mask. Following are the files used in this practical application:

– Red band: SENTINEL2A_20160928-105637-665_L2A_T31TCK_D_V1-0_FRE_B4.tif.

– Near-infrared band: SENTINEL2A_20160928-105637-665_L2A_T31TCK_D_V1-0_FRE_B8.tif.

– Mask: SENTINEL2A_20160928-105637-665_L2A_T31TCK_D_V1-0_MG2_R1.tif.

Link: https://theia.cnes.fr/atdistrib/rocket/#/search?collection=SENTINEL2

2.3.2. *Step 1: pre-processing*

Before applying the disaggregation method (based on the relationship between LST_{LR} and $NDVI_{LR}$), some pre-processing steps are needed: (1–3) obtaining the $NDVI_{HR}$, $NDVI_{LR}$ and LST_{LR} and applying the cloud mask to the S2 image, (4) subsetting the MODIS images to the same extent as the S2 image, (5) normalizing the two NDVI images (MODIS and S2). Steps 1–4 correspond to the first part of the organigram, and step 5 to the second part (Figure 2.1).

2.3.2.1. *Obtaining the NDVI and LST*

The first step is to obtain the NDVI for both HR and LR images (tasks 1 and 2 in first part of the organigram). Both NDVI images are obtained from the red and near-infrared bands using equation [2.2]. Typical NDVI values range between 0 and 1 for continental surfaces (soil and vegetation) and can be negatives, up to −1 for water.

Next the scale factor (0.02) is applied to the MOD11 image to obtain the LST image in K (LST LR). This is step 3 in the first part of the organigram.

Procedure	Manipulation in QGIS
1. Calculate the S2 NDVI image	In QGIS: • Open the two Sentinel-2 images (bands 4 and 8). In the **menu bar**: • Select **Raster > Raster Calculator…** In the **Raster calculator**: • Type the NDVI equation as: ("SENTINEL2A_20160928-105637-665_L2A_T31TCK_D_V1-0_FRE_B8@1" - "SENTINEL2A_20160928-105637-665_L2A_T31TCK_D_V1-0_FRE_B4@1") / ("SENTINEL2A_20160928-105637-665_L2A_T31TCK_D_V1-0_FRE_B4@1" + "SENTINEL2A_20160928-105637-665_L2A_T31TCK_D_V1-0_FRE_B8@1") • Name the output layer: NDVI_S2.TIF

2. Mask the pixels with water, clouds or shadows over the S2 image and the pixels with NDVI values greater than 1 or lower than 0	In QGIS: • Open the mask associated with the Sentinel-2 image: "SENTINEL2A_20160928-105637-665_L2A_T31TCK_D_V1-0_MG2_R1.tif" In the **menu bar**: • Select **Raster > Raster Calculator...** In the **Raster calculator**: • Type the equation: `"NDVI_S2@1" * ("SENTINEL2A_20160928-105637-665_L2A_T31TCK_D_V1-0_MG2_R1@1" != 1 AND "SENTINEL2A_20160928-105637-665_L2A_T31TCK_D_V1-0_MG2_R1@1" != 2 AND "SENTINEL2A_20160928-105637-665_L2A_T31TCK_D_V1-0_MG2_R1@1" != 8 AND "NDVI_S2@1" > 0 AND "NDVI_S2@1" < 1)` This equation allows us to keep only the pixels with values in the mask band different from 1 (water), 2 (clouds) and 8 (shadows), and the NDVI values greater than 0 and lower than 1. • Name the output layer: `NDVI_S2_masque.TIF`
3. Apply the same projection as S2 to the MODIS VNIR images	In QGIS: • Open the image MOD09GQ and select bands 1 and 2 ("sur_refl_b01_1" and "sur_refl_b02_1"). In the **menu bar**: • Select **Raster > Projections > Warp (Reproject)** In **Warp (Reproject)**: • Input file: select the image corresponding to band 1 • Output file: MODIS_b1.TIF • Target SRS: select EPSG:32631 that corresponds to UTM WGS84 zone 31 (this is the projection of S2 images) Repeat with band 2 (name it MODIS_b2.TIF)
4. Calculate the NDVI MODIS	In the **menu bar**: • Select **Raster > Raster Calculator...** In the **Raster Calculator**: • Type the equation for the NDVI. The conditions for masking pixels with values greater than 0 or lower than 1 can be added here with a logic operation. The equation is applied only to the pixels for which the result is greater than 1 or lower than 0. Pixels that do not fulfill this requirement will remain at 0. This avoids having to undertake a second step in order to mask these pixels.

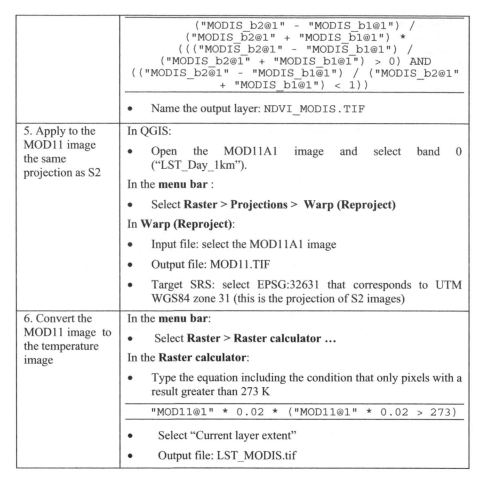

	("MODIS_b2@1" - "MODIS_b1@1") / ("MODIS_b2@1" + "MODIS_b1@1") * ((("MODIS_b2@1" - "MODIS_b1@1") / ("MODIS_b2@1" + "MODIS_b1@1") > 0) AND (("MODIS_b2@1" - "MODIS_b1@1") / ("MODIS_b2@1" + "MODIS_b1@1") < 1)) • Name the output layer: NDVI_MODIS.TIF
5. Apply to the MOD11 image the same projection as S2	In QGIS: • Open the MOD11A1 image and select band 0 ("LST_Day_1km"). In the **menu bar** : • Select **Raster** > **Projections** > **Warp (Reproject)** In **Warp (Reproject)**: • Input file: select the MOD11A1 image • Output file: MOD11.TIF • Target SRS: select EPSG:32631 that corresponds to UTM WGS84 zone 31 (this is the projection of S2 images)
6. Convert the MOD11 image to the temperature image	In the **menu bar**: • Select **Raster** > **Raster calculator ...** In the **Raster calculator**: • Type the equation including the condition that only pixels with a result greater than 273 K "MOD11@1" * 0.02 * ("MOD11@1" * 0.02 > 273) • Select "Current layer extent" • Output file: LST_MODIS.tif

Table 2.1. *Obtaining the NDVI and LST*

MODIS images (*NDVI_MODIS*, *LST_MODIS*) have a larger spatial coverage than S2 images (*NDVI_S2*). They also have a lower spatial resolution. It is necessary that all images cover the same extent and that they have the same number of pixels when they are resized from HR to LR and vice versa. To accomplish these conditions, the "align" function from QGIS can be used. This function subsets and aligns the images, giving them the same spatial resolution. The LR images are first resized to the HR (using the nearest neighbor approach), and they will be later resized to its original spatial resolution (250 m for $NDVI_{LR}$, and 1,000 m for LST_{LR}). It is also necessary to verify that the dimensions of the LR are a multiple number of the dimensions of the HR image so that when the HR images are resized to the LR, the resulting image will have a whole number of lines and columns.

The subsetting of the images corresponds to task 4 in the first part of the organigram. This task is applied to all images (HR and LR).

Procedure	Manipulation in QGIS
1. Subsetting	In the **menu bar**: • Select **Raster > Align Rasters…** In **Align Rasters**: • Add the images LST_MODIS, NDVI_MODIS and NDVI_S2_masque. For each image give the corresponding output name: LST_sub10m.tif, NDVI_MODIS_sub10m.tif and NDVI_HR.tif • In "Reference Layer" select the NDVI_S2_masque image • Select the option "Click to Extent", next select "Layer extent" and we should see the following coordinates and the Output Size 10980 × 10980: 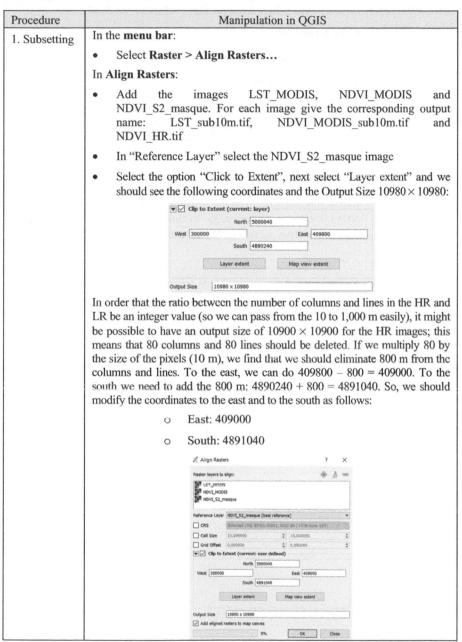 In order that the ratio between the number of columns and lines in the HR and LR be an integer value (so we can pass from the 10 to 1,000 m easily), it might be possible to have an output size of 10900 × 10900 for the HR images; this means that 80 columns and 80 lines should be deleted. If we multiply 80 by the size of the pixels (10 m), we find that we should eliminate 800 m from the columns and lines. To the east, we can do 409800 − 800 = 409000. To the south we need to add the 800 m: 4890240 + 800 = 4891040. So, we should modify the coordinates to the east and to the south as follows: ○ East: 409000 ○ South: 4891040

2. Consider the negative values in the $NDVI_{LR}$, $NDVI_{HR}$ and LST_{LR} images as no data	The function **Align Rasters** replaces the no data pixels in the LST_sub10m, NDVI_MODIS_sub10m and NDVI_HR images by negative values. In order to put these values as "no data": In the **menu bar**: • Select **Raster > Raster Calculator...** In the **Raster Calculator**: • Type the following equation to define negative values as "no data" $$(-1/(("NDVI\_HR@1"<=0)-1))*"NDVI\_HR@1"$$ Select "Current layer extent": • Output file: NDVI_HR_na.tif Repeat this step to convert the negative values in the LST_sub10m and NDVI_MODIS_sub10m images to "no data". Name the output images as: LST_sub10m_na.tif, and NDVI_MODIS_sub10m_na.tif. 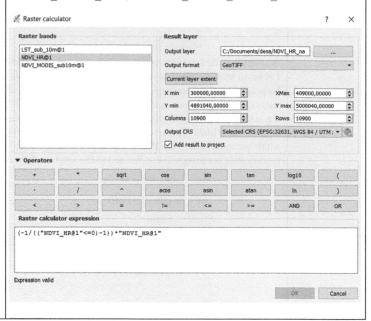

3. Resample the MODIS images to their original resolutions	The output images from the previous step (LST_sub10m_na, NDVI_MODIS_sub10m_na) have a spatial resolution equivalent to the reference image (10 m). The MODIS NDVI and LST images will now be resampled to their original resolution (250 and 1,000 m, respectively).
	To do the resampling, it is necessary for "Processing" to be activated and visible in the menu bar. If it is not activated, it is necessary to activate it: Plugins > Manage and install plugins: search for "Processing".
	In the **menu bar**:
	• Select **Processing** > **Toolbox**
	In the **Toolbox**:
	• Search "**r.resamp.stats**": GRASS GIS 7 comands > Raster > r.resamp.stats
	In **r.resamp.stats**:
	• In **Input raster layer** select the LST_sub10m_na image
	• In **Aggregation method** select **average**
	• In **GRASS GIS 7 region cellsize...** type 1000
	• In **Resampled aggregated** type the name of the output image: LST_LR_na.tif
	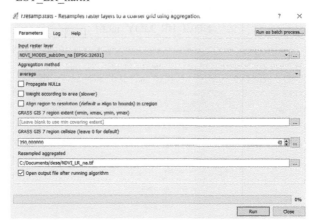
	Repeat with the NDVI_MODIS_sub10m_na image giving it the output name "NDVI_LR_na.tif" and the region cellsize 250.

Table 2.2. *Cutting images*

2.3.2.2. *Normalization of the NDVI images*

The information captured by the red and near-infrared bands from two different sensors may be slightly different, especially due to the different spectral resolution of the bands in both sensors but also to the different bandwidth. A relationship can be obtained between the NDVI at LR from the two sensors (the $NDVI_{HR}$ from S2 needs to be resampled to the spatial resolution of the MODIS NDVI). Next, the relationship can be applied to the $NDVI_{HR}$ image. This step corresponds to the second part of the organigram (Figure 2.1).

Procedure	Manipulation in QGIS
1. Resample the NDVI_HR_na image	In the **menu bar**: • Select **Processing > Toolbox** In the **Toolbox**: • Search "**r.resamp.stats**": **GRASS GIS 7 commands > Raster > r.resamp.stats** In **r.resamp.stats**: • In **Input raster layer** select the NDVI_HR_na image • In **Aggregation method** select **average** • In **GRASS GIS 7 region cellsize…** type 250 • In **Resampled aggregated** type the name of the output image: NDVI_HR_MOY_na.tif

2. Obtain the linear regression between NDVI_LR and NDVI_HR_MOY _na	In the **menu bar**: • Select **Processing > Toolbox…** In the **toolbox**: • Search "regression": **GRASS GIS 7 commands > Raster > r.regression.line** In **r.regression.line**: • Layer for y coefficient: select "NDVI_LR_na" • Layer for x coefficient: select "NDVI_HR_MOY_na" • Select "Run" • Result: 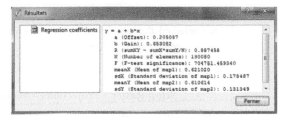 • Write down the coefficients "a1" and "b1"
3. Apply the previous equation to the NDVI_HR image	In the **menu bar:** • Select **Raster > Raster Calculator…** In **Raster Calculator**: • Type the equation a1+b1*x, where "a1" and "b1" are the coefficients obtained in the previous step (a1 = 0.21 and b1 = 0.65 in this example). "x" is the NDVI_HR_na image. Add the conditions for masking values lower than 0 and higher than 1. These conditions can be integrated in the equation by adding a logical operation, multiplying the result by 1 when the condition "((a1+b1*x)>0) & ((a1+b1*x)<1)" is true. Pixels where this is false are set to 0. This avoids making an extra step for masking pixels with values greater than 1 or lower than 0. (0.21 + 0.65 * "NDVI_HR_na@1") * (((0.21 + 0.65 * "NDVI_HR_na@1")>0) AND ((0.21 + 0.65 * "NDVI_HR_na@1")<1)) • Select the NDVI_HR_na image and "Current layer extent" • Output layer: NDVI_HR_norm.tif

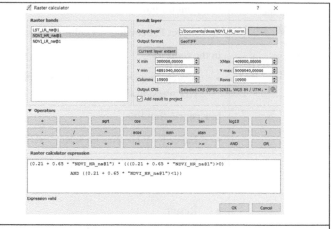

4. Define the 0 as no data	In the previous step, we have left to 0 the pixels that did not accomplished the condition. Now, we should specify that this "0" values correspond to "no data".
	In the **menu bar**:
	• Select **Raster > Projections > Warp (Reproject)...**
	In **Warp (Reproject)**:
	• Input file: NDVI_HR_norm
	• Output file: NDVI_HR_norm_na.tif
	• No data values: 0

Table 2.3. *Normalization of NDVI images*

2.3.3. *Step 2: disaggregation*

2.3.3.1. *First simulation of the LST HR*

The first part of the disaggregation method has several tasks:

– Obtain the variation coefficient (CV) that will be used to select the most homogeneous pixels in the $NDVI_{LR}$ image.

– Obtain the linear regression $NDVI_{LR}$ and LST_{LR} using homogenous pixels.

– Apply the equation obtained in task 2 to the image $NDVI_{HRnorm}$ (to obtain LST'_{HR}).

This first part of the disaggregation corresponds to part 3 of the organigram (Figure 2.1).

2.3.3.1.1. Variation coefficient image

The variation coefficient image allows us to select the most homogeneous pixels in the $NDVI_{LR}$ image (1 km). These pixels will be used to obtain the linear regression between $NDVI_{LR}$ and LST_{LR}. The variation coefficient is obtained for each pixel in the $NDVI_{LR}$ image using all the 10 m resolution pixels from the $NDVI_{HR}$ belonging to each 1 km pixel of the $NDVI_{LR}$. The variation coefficient is obtained as the ratio between the standard deviation and the average value of the 10 m pixels (equation [2.5]). This task corresponds to task 1 in the third part of the organigram (Figure 2.1).

Procedure	Manipulation in QGIS
1. Average the $NDVI_{HR}$ at 1 km	In the **menu bar**: • Select **Processing** > **Toolbox** In the **Toolbox**: • Search "**r.resamp.stats**": **GRASS GIS 7 commands** > **Raster** > **r.resamp.stats** In **r.resamp.stats**: • In **Input raster layer** select the NDVI_HR_norm_na image • In **Aggregation method** select **average** • In **GRASS GIS 7 region cellsize…** type 1000

	• In **Resampled aggregated** type the name of the output image: AvL.tif 
2. Resample the previous image to 10 m	In the **menu bar**: • Select **Processing** > **Toolbox** In the **Toolbox**: • Search "**r.resamp.interp**": GRASS GIS 7 commands > **Raster** > **r.resamp.interp** In **r.resamp.interp**: • In **Input raster layer** select the AvL image • In **Sampling interpolation method** select **nearest** • In **GRASS GIS 7 region cellsize…** type 10 • In **Resampled aggregated** type the name of the output image: AvH.tif 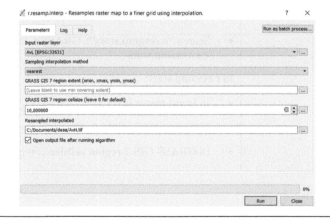

3. Difference image between each $NDVI_{HR}$ and the corresponding NDVI averaged at 1 km (AvL)	In the **menu bar:** • Select **Raster > Raster Calculator…** In **Raster Calculator**: • Type the equation: $$(\text{"NDVI\_HR\_norm\_na@1"} - \text{"AvH@1"})\char`^2$$ • Select the image NDVI_HR_norm_na and select "Current layer extent" • Output file: Dev_na.tif 
4. Variance image of the resampled $NDVI_{HR}$ at 1 km	Repeat the first task of this section to resample the Dev_na image to 1,000 m. Output name: AvDev.tif 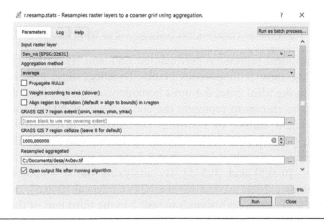

5. Variation coefficient	In the **menu bar**: • Select **Raster > Raster Calculator…** In **Raster Calculator**: • Type the equation: $$\text{sqrt("AvDev@1") / "AvL@1"}$$ • Select the image MoyL and "Current layer extent" • Output file: CV.tif 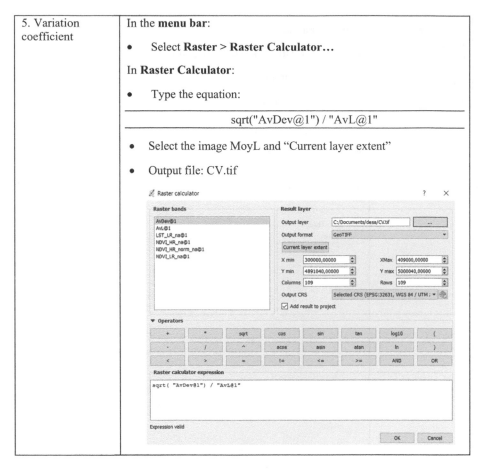

Table 2.4. *Variation coefficient computation*

2.3.3.1.2. Linear regression between the temperature (LST) and NDVI

The linear regression between the LST and NDVI images should be completed using the most homogeneous pixels. To select these pixels, the CV image can be used. In the ideal case, the most homogeneous pixels (e.g. the 20% of these pixels) should be chosen to represent the entire range of NDVI values (between 0 and 1). This would consist of choosing homogenous pixels from different intervals of NDVI: [0–0.2], [0.2–0.5] and [0.5–1]. In the image used in this chapter, there are few pixels with values lower than 0.2. So, in this case only two intervals have been chosen: [0–0.5] and [0.5–1]. By analyzing the CV values in both intervals, the thresholds chosen are 0.3 for the NDVI interval [0–0.5], and 0.15 for the interval [0.5–1]. This task corresponds to task 2 in the third part of the organigram.

Procedure	Manipulation in QGIS
1. Resample the MODIS NDVI image to 1 km	Repeat task number 1 in the preceding section to resample the NDVI_LR_na image to 1,000 m. Output image: NDVI_LR_Av.tif. 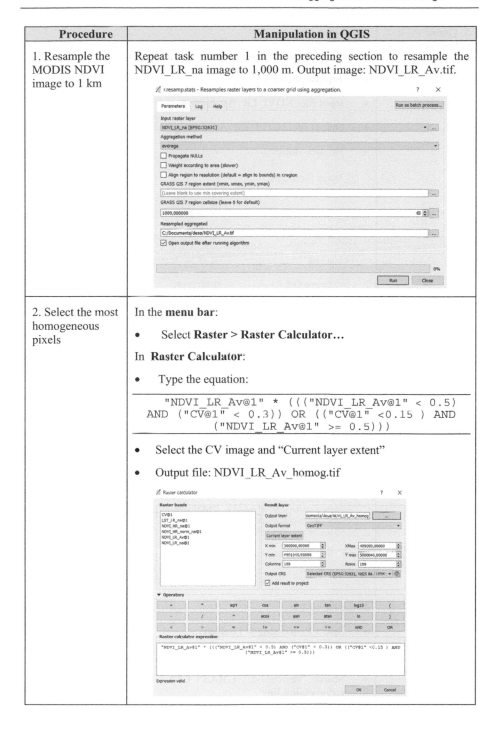
2. Select the most homogeneous pixels	In the **menu bar**: • Select **Raster > Raster Calculator…** In **Raster Calculator**: • Type the equation: `"NDVI_LR_Av@1" * ((("NDVI_LR_Av@1" < 0.5) AND ("CV@1" < 0.3)) OR (("CV@1" <0.15 ) AND ("NDVI_LR_Av@1" >= 0.5)))` • Select the CV image and "Current layer extent" • Output file: NDVI_LR_Av_homog.tif

3. Define the 0 values as no data in the image NDVI_LR_Av_homog	In the **menu bar**: • Select **Raster > Projections > Warp (Reproject)…** In **Warp (Reproject)**: • Input file: NDVI_LR_Av_homog • Output file: NDVI_LR_Av_homog _na.tif • No data values: 0 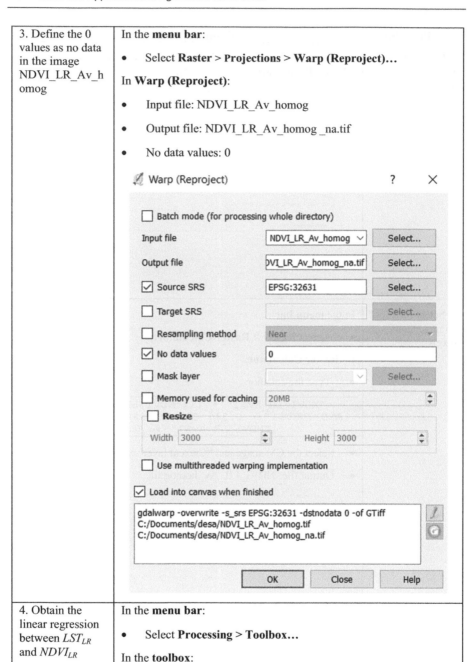
4. Obtain the linear regression between LST_{LR} and $NDVI_{LR}$	In the **menu bar**: • Select **Processing > Toolbox…** In the **toolbox**: • Search "regression": **GRASS GIS 7 commands > Raster > r.regression.line**

In **r.regression.line**:

- Layer for x coefficient: select "NDVI_LR_Av_homog_na"

- Layer for y coefficient: select "LST_LR _na"

- Select "Run"

- Result:

- Write down the coefficients "a" and "b"

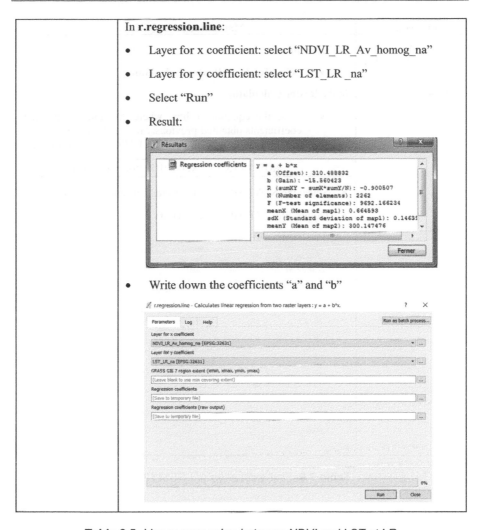

Table 2.5. *Linear regression between NDVI and LST at LR*

2.3.3.1.3. First estimation of LST HR

The previous equation obtained from the linear regression between LST_{LR} and $NDVI_{LR}$ is now applied to the NDVI HR normalized image ($NDVI_{HRnorm}$, equation [2.4]) to obtain a first estimation of the simulated LST HR. This task corresponds to task 3 of the third part of the organigram (Figure 2.1).

Procedure	Manipulation in QGIS

1. Apply the equation obtained at the end of section 2.3.2.2	In the **menu bar**:
	
	   • Select **Raster > Raster Calculator…**
	
	In the **Raster Calculator**:
	
	   • Type the equation a+b*x, where "a" and "b" are the coefficients obtained previously (a = 310.5 and b = −15.5 in this example). "x" is the NDVI_HR_norm_na image.
	
	     (310.5 - 15.5 * "NDVI_HR_norm_na@1")
	
	   • Select the NDVI_HR_norm_na and "Current layer extent"
	
	   • Output file: LST_HR_prime_na.tif
	
	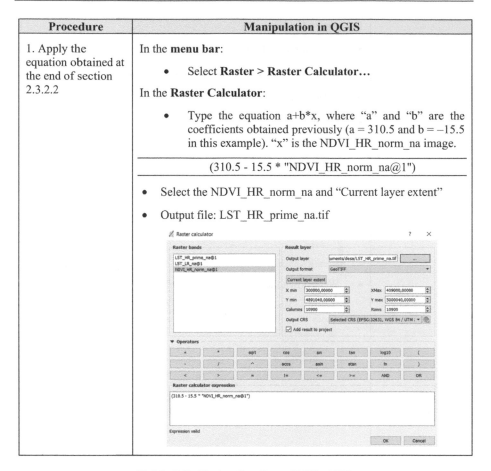

Table 2.6. *First estimation of LST at HR*

2.3.3.2. *Residual correction*

The equation obtained between LST_{LR} and $NDVI_{LR}$ from the most homogeneous pixels of the $NDVI_{LR}$ image does not consider the possible local effects. These effects can be somehow corrected by applying a residual correction. This includes the following tasks:

– The first simulated LST image (LST'_{HR}) is first resampled to the LR (LST_{HR}^{Av}).

– A residual image (Residual) is obtained as the difference between the LST_{HR}^{Av} image and the actual MODIS LST image (LST_{LR}).

– The residual image is resampled to 10 m spatial resolution and smoothed using a Gaussian filter ($Residual_{HRsm}$).

– The smoothed residual image is subtracted from the first estimation of the LST HR (LST'$_{HR}$) to obtain the temperature at HR (LST$_{HR}$).

These four tasks correspond to those in the fourth part of the organigram (Figure 2.1).

Procedure	Manipulation in QGIS
1. Resample the simulated LST image (*LST'$_{HR}$*) to 1 km	Repeat the step 1 in section 2.3.3.1.1 to resample the LST_HR_prime_na to 1,000 m. Output file: LST_HR_Av.tif. 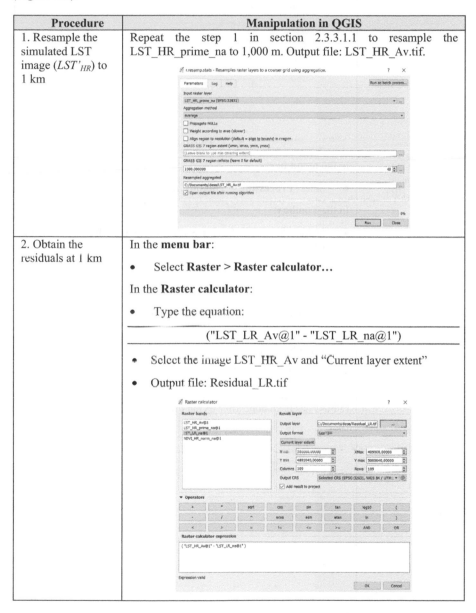
2. Obtain the residuals at 1 km	In the **menu bar**: • Select **Raster > Raster calculator…** In the **Raster calculator**: • Type the equation: <hr>("LST_LR_Av@1" - "LST_LR_na@1")<hr>• Select the image LST_HR_Av and "Current layer extent" • Output file: Residual_LR.tif

3. Resample the residuals to 10 m	Repeat task 2 in section 2.3.3.1.1 to resample the Residual_LR image to 10 m. Output file: Residual_HR.tif. 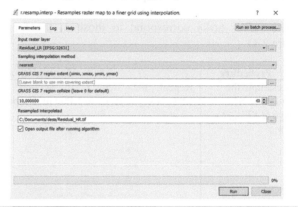
4. Smooth the residual image (Residual_HR)	In the **menu bar**: • Select **Processing > Toolbox…** In the **Toolbox**: • Search "Gaussian filter": **SAGA > Raster filter > Gaussian filter** In **Gaussian filter**: • Grid: select the image "Residual_HR" • Standard Deviation (controls the intensity of the smoothing): 3 • Search Radius (size of the pixels group to be used in the smoothing): 15 • Filtered Grid: residual_HRsm The choice of values for "standard deviation" and "search radius" has been done after trying different values. 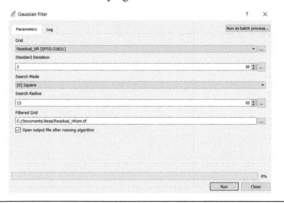

| 5. Subtract the smoothed residuals (Residual_HRsm) from the simulated LST image (LST'_{HR}) | In the **menu bar**:

• Select **Raster > Raster Calculator…**

In the **Raster Calculator**:

• Type the equation:

`("LST_HR_prime_na@1" - "Residual_HRsm@1")`

• Select the LST_HR_prime_na image and "Current layer extent"

• Output file: LST_HR.tif

 |

Table 2.7. *Residual correction*

2.4. Results analysis

The disaggregation method of thermal images applied to images from different sensors (MODIS and S2) allows for the simulation of a thermal image at 10 m spatial resolution. S2 provides images with a temporal resolution of 5 days, and MODIS daily, so it will be possible to have disaggregated thermal images with a spatial resolution of 10 m and temporal resolution of 5 days.

Figure 2.3 shows the LST images with spatial resolutions of 10 m (disaggregated image: LST_{HR}) and 1,000 m (MODIS image: LST_{LR}). Figure 2.3(b) shows how the LST_{HR} image allows us to identify more details from the surface, especially parcels, and presents a wider range of LST values than the MODIS image (LST_{LR}). A 1,000 m pixel may cover a variety of objects having high and low LST values, this results in an average LST value for that pixels.

The disaggregation method has been evaluated over pairs of MODIS-Landsat images [BIS 16a], and MODIS-S2 images [BIS 16b]. Errors (root mean square error (RMSE)) of about 2 K have been observed for the entire images. However, large errors may be present in some cases in the disaggregated images. The largest errors are observed in highly heterogeneous agricultural areas, and especially in small parcels that have been irrigated just before the thermal image has been acquired. When a parcel has been irrigated, the LST decreases but the effect over the NDVI (LR or HR) is not observed immediately. So, when applying the equation obtained from the NDVI and LST, the resulting LST will be higher than the real one. Moreover, if the parcel is small compared to the LR pixel, the effect of the irrigation is not captured in the MODIS LST image due to its low resolution, so even by applying the residual correction there will always be this error in the LST estimation.

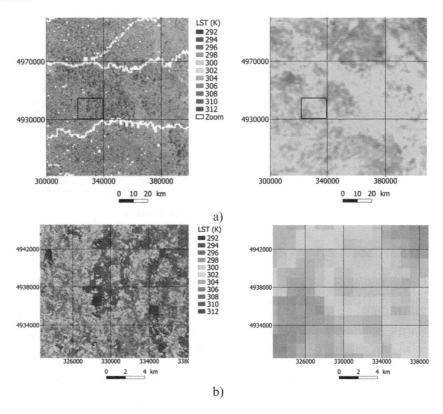

Figure 2.3. *a) LST images: disaggregated (left, 10 m) and MODIS (right, 1,000 m), and b) zoom into the images from (a) corresponding to the black square. The coordinates system is UTM WGS84 zone 31N. For a color version of the figure, see www.iste.co.uk/baghdadi/qgis2.zip*

2.5. Bibliography

[AGA 07] AGAM N., KUSTAS W.P., ANDERSON M.C. *et al.*, "A vegetation index based technique for spatial sharpening of thermal imagery", *Remote Sensing of Environment*, vol. 107, no. 4, pp. 545–558, 2007.

[BIN 13] BINDHU V.M., NARASIMHAN B., SUDHEER K.P., "Development and verification of a non-linear disaggregation method (NL-DisTrad) to downscale MODIS land surface temperature to the spatial scale of Landsat thermal data to estimate evapotranspiration", *Remote Sensing of Environment*, vol. 135, pp. 118–129, 2013.

[BIS 15] BISQUERT M., BORDOGNA G., BÉGUÉ A. *et al.*, "A simple fusion method for image time series based on the estimation of image temporal validity", *Remote Sensing*, vol. 7, no. 1, pp. 704–724, 2015.

[BIS 16a] BISQUERT M., SÁNCHEZ J.M., CASELLES V., "Evaluation of disaggregation methods for downscaling MODIS land surface temperature to Landsat spatial resolution in Barrax test site", *IEEE Journal of Selected Topics in Applied Earth Observations and Remote Sensing*, vol. 9, no. 4, pp. 1430–1438, 2016.

[BIS 16b] BISQUERT M., SÁNCHEZ J.M., LÓPEZ-URREA R. *et al.*, "Estimating high resolution evapotranspiration from disaggregated thermal images", *Remote Sensing of Environment*, vol. 187, pp. 423–433, 2016.

[GAO 06] GAO F., MASEK J., SCHWALLER M. *et al.*, "On the blending of the Landsat and MODIS surface reflectance: predicting daily Landsat surface reflectance", *IEEE Transactions on Geoscience and Remote Sensing*, vol. 44, no. 8, pp. 2207–2218, 2006.

[GAO 12] GAO F., KUSTAS W., ANDERSON M., "A data mining approach for sharpening thermal satellite imagery over land", *Remote Sensing*, vol. 4, no. 12, pp. 3287–3319, 2012.

Automatic Extraction of Agricultural Parcels from Remote Sensing Images and the RPG Database with QGIS/OTB

3.1. Context

Within the framework of the common agricultural policy (CAP), farmers must declare the nature and surfaces of their farm crops for which they apply to the European Union for aid. Since 2002, these declarations have been gathered in a database, which became cartographic in 2006 and which is called the RPG (*Registre Parcellaire Graphique*) in France. The RPG is the French version of the European *Land Parcel Identification System* (LPIS) database [MAR 14]. The RPG, managed by the French Agency of Services and Payment (ASP), is on a 1/5,000 scale and makes it possible to precisely locate the main crops at the national territory level each year. This information makes it possible to study the geographical distribution of crops and, by combining several years of RPG, to reconstitute crop sequences, which comprises key information on the farmers' practices and their potential impacts on the environment. RPG data are organized around "crop groups". Each crop group includes one or several individual crops. The Agricultural Pressure and Socio-Economic Coordination on water catchment areas (PACS-AAC) project, set up by the INRA/AgroParisTech SADAPT research unit and financed by the French agency for biodiversity, aims at spatially rebuilding crop sequences on a territory around a catchment area, in order to carry out better diagnosis and monitoring for water resources protection. SADAPT thus developed RPG explorer software

Chapter written by Jean-Marc GILLIOT, Camille LE PRIOL, Emmanuelle VAUDOUR and Philippe MARTIN.

[MAR 17] in order to automate the reconstitution of crop sequences from several years of RPG data. However, until 2015, a limit associated with RPG data was that crop information was not precisely located at the agricultural plot (AP) level, but related to farmer's block (FB)/"îlot" level [SAG 08]. Referring to the spatial arrangement of crops, it is indeed necessary to distinguish different possible geographical entities:

– Cadastral parcel (CP) is the geographical entity for agricultural ownership.

– Agricultural plot (AP) a geographical entity that refers to agricultural activity in which only one crop is grown, which can be either a subset of one CP or a grouping of several CPs.

– Farmer's block (FB) or "îlot" is the geographical entity in the RPG data for crop groups, which is defined as one or several contiguous APs managed by the same farmer.

Figure 3.1. *FB parcels on an SPOT-5 satellite ortho-image background (May 7, 2008) in infrared colored composite, near Avremesnil city (49.859°N , 0.917°E) in the Seine-Maritime department, France. ©CNES 2008, distribution Airbus DS/Spot Image. For a color version of the figure, see www.iste.co.uk/baghdadi/qgis2.zip*

Figure 3.1 shows an example of a FB containing only one AP (4282403) and another, containing two APs (4325322). The objective of the work proposed in this chapter is to implement a processing method, by means of the QGIS software, combining the RPG data and satellite images of the same year, in order to automatically extract and inform each AP inside a given FB. This information should improve the quality of the crop sequences produced by RPG explorer. In order to extract the AP limits, the Orfeo ToolBox (OTB) image processing library developed by the French space agency (CNES) and integrated in QGIS will be used. In order to be able to easily use this method on different RPG datasets, we will implement it as a Python user script, in QGIS [LEP 16].

3.2. Method of AP extraction

The method of AP extraction is subdivided into three main steps (Figure 3.2):

– formatting of the RPG data;

– classification of satellite images;

– intersect overlay between extracted AP and FB, then crop validation.

3.2.1. *Formatting the RPG data*

This first step (Figure 3.2) is pre-processing of the RPG data in order to group information, which is initially distributed between several files, into a single layer, making it easier to manipulate for the next processing steps. A projection of the layers into the French official geographic system, RGF93 Lambert 93, is first applied. Having the layers in the same projection system guarantees better precision with spatial analysis functions under GIS. FBs that are in the study area are extracted with a spatial join. Then a Python language instruction loops into the RPG information to separate and join each crop group data to its FB. Finally, FBs containing only one group of crops, which are called "pure FB", are separated from those that have several, known as "poly FB". An FB layer version is produced by applying a buffer zone of -10 m, so as to eliminate edge effects at the limits between parcels for the following analyses.

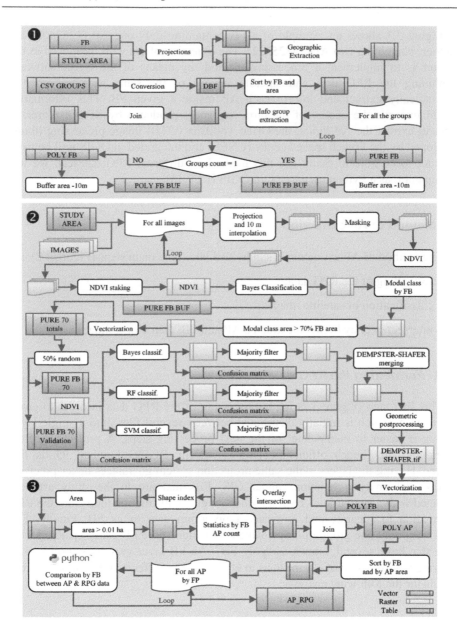

Figure 3.2. *Processing workflow for agricultural plot extraction. For a color version of the figure, see www.iste.co.uk/baghdadi/qgis2.zip*

3.2.2. *Classification of SPOT satellite images*

The second step (Figure 3.2) deals with the processing of satellite images. We used a time series of images over the studied growing period. Images were provided corrected in apparent reflectance at the top of the atmosphere (TOA). To characterize the crop development in the parcels and thus map them, we calculated the most commonly used vegetation index, the standardized vegetation index or *NDVI* (Normalized Difference Vegetation Index):

$$NDVI = \frac{R_{nir} - R_r}{R_{nir} + R_r} \qquad\qquad [3.1]$$

where R_{nir} is reflectance in the near-infrared channel and R_r is reflectance in the red one. *NDVI* is, under certain conditions, directly related to the development of plant biomass; different *NDVI* values are likely to discriminate between plots and thus make it possible to map them. In some cases, different crops may have the same appearance at specific stages of development, therefore several images acquired at different dates during the growing season are used for an improved discrimination between crops. Such a temporal approach is called a diachronic analysis. For each date a new *NDVI* band is computed, and then all these bands are stacked together in the same composite image, which is then classified. Half of the pure FBs are used as learning areas or ROIs (Region of Interest) to calculate the reference statistics for this classification, while the other half is used to calculate a confusion matrix for validation. This kind of classification, based on references provided by the operator, is called an assisted or supervised classification. In order to improve performance, three different supervised classifications are performed: *Bayes*, *Random Forest* (RF) and *Support Vector Machine* (SVM). A final merging or "fusion" step of the three classifications produces a more robust single classification than the original ones. The method of fusion typically relies on the confusion matrices calculated during the three initial classifications, in order to calculate class likelihood functions, and then in each pixel, the class with the maximum likelihood is kept as the final class. The Dempster–Shafer fusion method was used [SHA 76]. Post-processing was applied to the classified images in order to cartographically smooth the results and also to mask the noise due to edge effects between the FB limits.

3.2.3. *Intersect overlay between extracted AP and FB with crop validation*

In this last step (Figure 3.2), the candidate APs are produced in each FB. The final classified image is first vectorized and then combined with the poly FBs by an overlay intersection operation, in order to:

– limit the resulting parcels to the FB shape;

– "retrieve" their RPG information.

Finally, for each plot, its area is computed in order to compare it with the crop areas in the RPG data, and a shape index is also calculated (equation [3.2]) to eliminate very irregularly shaped artifact polygons:

$$ShapeIndex = \frac{perimeter}{2\sqrt{\pi\ area}}$$ [3.2]

Areal filtering is applied to eliminate tiny polygons (area ≤ 0.01 ha) that are processing artifacts. Counting is carried out using the "statistics by category" function, in order to determine, for each FB, the number of APs and groups of crops. The result layer is finally sorted by FB and by decreasing area. A Python loop will iterate over the result layer to analyze the obtained APs in each FB to validate, delete, reassign or merge them, relying on the crops' RPG data.

3.3. Practical application of the AP extraction

This section presents a practical application of the AP extraction from the RPG and SPOT satellite images on the Dun catchment area in the Seine-Maritime department (Figure 3.3) for the 2007–2008 growing season. The overall extent of the study is the boundary of the Dun basin (northwestern France), which covers an area of 10,857 ha.

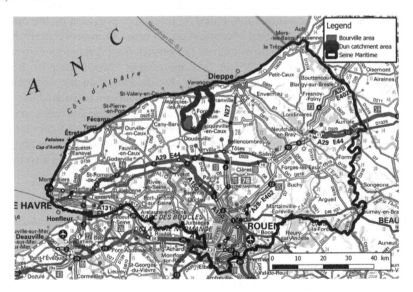

Figure 3.3. *Situation map of the Dun catchment (49.816°N, 0.873°E) in the department of Seine-Maritime, France, background IGN-SCAN1000 (see: www.ign.fr). For a color version of the figure, see www.iste.co.uk/baghdadi/qgis2.zip*

In order to finalize the procedure, we worked on a small area of 1,480 ha: the Bourville catchment area (Figure 3.3), since a crops map was available for 2008, and could be used as validation data.

3.3.1. *Software and data*

3.3.1.1. *Required software*

The QGIS version used is 2.14.3 Essen 32 bits with Grass, under Windows 10. The OTB image processing library developed by CNES is also used.

3.3.1.2. *Input data*

3.3.1.2.1. RPG data

The RPG data used are those provided by the ASP. A simplified version of the RPG is also available as opendata on www.data.gouv.fr, but this version only contains the main crop group per FB. Since 2017, the National Institute of Geographic and Forest Information (IGN) has been responsible for the distribution of anonymized RPG data, which are now provided under open license, that is free of charge for all uses, including commercial ones. The shipped RPG data consist of a GIS file in ShapeFile format, which gives the outline of the FB, and files in Comma Separated Value (CSV) format, which describe their content, especially "ilot_groupe_culture.csv", which gives the crops present in the FB (Table 3.1).

CODE	CROP	CODE	CROP
0	NO INFORMATION	15	GRAIN LEGUMES
1	SOFT WHEAT	16	FODDER
2	MAIZE GRAINS AND SILAGE	17	PASTURE LANDS
3	BARLEY	18	PERMANENT GRASSLAND
4	OTHER CEREALS	19	TEMPORARY GRASSLAND
5	RAPESEED	20	ORCHARDS
6	SUNFLOWER	21	VINEYARDS
7	OTHER OILSEEDS	22	NUTS
8	PROTEIN CROPS	23	OLIVE TREES
9	FIBRE PLANTS	24	OTHER INDUSTRIAL CROPS
10	SEEDS	25	VEGETABLES-FLOWERS
11	SET-ASIDE AREAS NO PRODUCTION	26	SUGAR CANE
12	INDUSTRIAL SET-ASIDE AREAS	27	FRUIT TREE
13	OTHER SET-ASIDE AREAS	28	VARIOUS
14	RICE		

Table 3.1. *RPG crop groups coding*

In the "ilot_groupe_culture.csv" file, the following fields are found: "ID_ILOT", a unique identifier of the FB, "CODE_GROUPE_CULTURE", crop group code according to the codification in Table 3.1 and "SURFACE_GROUPE_CULTURE", group area in hectares within the FB.

3.3.1.2.2. Download of the *SPOT World Heritage program* images

This exercise uses SPOT images provided by the SPOT World Heritage (SWH) program. SWH is a program from the CNES that provides SPOT satellites images over 5 years of age, free of charge, for non-commercial uses. Data are available to all registered users of the "Theia" continental services and data center Website at the following link: https://www.theia-land.fr/fr/produits/spot-world-heritage (Figure 3.4).

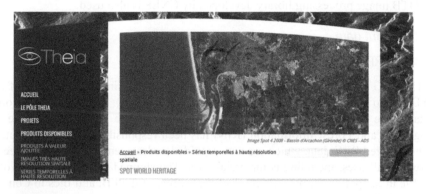

Figure 3.4. *Theia Website for downloading SPOT World Heritage images. For a color version of the figure, see www.iste.co.uk/baghdadi/qgis2.zip*

In order to search for candidate images, a graphical interface on the Theia site allows the image database to be queried according to date, as well as geographical and processing level criteria (Figure 3.5). Small images or "quicklook" display an overview of the result images. Care is taken to avoid images that are too cloudy.

				Accueil	Aide	Se connecter	S'enregistrer	
France	12 aout 2008 - 10:47:37	SPOT4	HRVIR1	REFLECTANCETOA	0 m		LEVEL1C	XS
France	06 aout 2008 - 11:03:10	SPOT4	HRVIR1	REFLECTANCETOA	0 m		LEVEL1C	XS
France	30 juillet 2008 - 11:05:57	SPOT5	HRG2	REFLECTANCETOA	0 m		LEVEL1C	XS
France	01 juillet 2008 - 11:17:34	SPOT2	HRV1	REFLECTANCETOA	0 m		LEVEL1C	XS

Figure 3.5. *Query result example on the Theia SPOT World Heritage Website. For a color version of the figure, see www.iste.co.uk/baghdadi/qgis2.zip*

Images are selected covering the entire study area, showing little or no clouds and throughout the crop period. For the Bourville watershed, five images were selected from December 19, 2007 to September 27, 2008 (Figure 3.6).

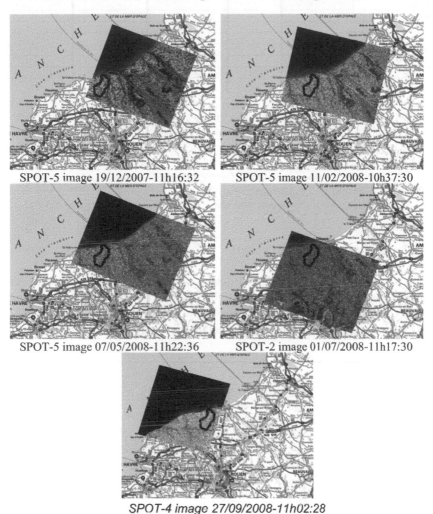

SPOT-4 image 27/09/2008-11h02:28

Figure 3.6. *The five selected SPOT images ©CNES 2007 and 2008, distribution Airbus DS/Spot Image, background IGN-SCAN1000 (see: www.ign.fr). For a color version of the figure, see www.iste.co.uk/baghdadi/qgis2.zip*

The choice of image dates seeks to maximize the differences between the main crops present in the study area, so we must also characterize the typical crop calendar of the Normandy region (Figure 3.7).

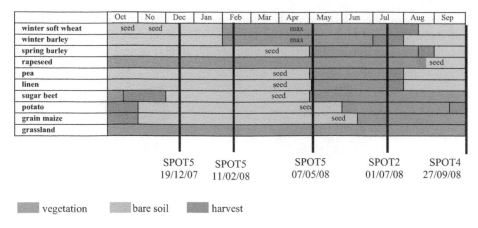

	Oct	No	Dec	Jan	Feb	Mar	Apr	May	Jun	Jul	Aug	Sep
winter soft wheat	seed	seed					max					
winter barley							max					
spring barley						seed						
rapeseed												seed
pea						seed						
linen						seed						
sugar beet						seed						
potato							seed					
grain maize								seed				
grassland												

SPOT5	SPOT5	SPOT5	SPOT2	SPOT4
19/12/07	11/02/08	07/05/08	01/07/08	27/09/08

vegetation   bare soil   harvest

Figure 3.7. *Temporal position of selected SPOT images, on a typical crop calendar in the Normandy region (adapted from [LEP 16]). For a color version of the figure, see www.iste.co.uk/baghdadi/qgis2.zip*

3.3.2. *Setting up the Python script*

This section presents the technical details of the development of a user script, which can be reproduced by the reader of this chapter. User scripts can be integrated into the QGIS processing toolbox, to display it: Menu → Processing → ToolBox. The "Processing Toolbox" pane appears on the right side of the screen (Figure 3.8).

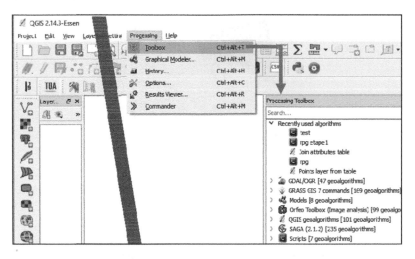

Figure 3.8. *Display of the QGIS "Processing Toolbox" pane, from the "Processing" item of the general menu bar. For a color version of the figure, see www.iste.co.uk/baghdadi/qgis2.zip*

We create a new Python script for our application, from the toolbox:

 Scripts ➜ Tools ➜  Create new script

The "Script Editor" window appears (Figure 3.9). You must then enter a comment line (starting with a "#" character) and then save the file, for example under the name rpg.py, with the button 🖫 .

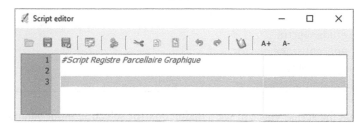

Figure 3.9. *The QGIS "Script Editor" window. For a color version of the figure, see www.iste.co.uk/baghdadi/qgis2.zip*

By default, scripts are saved in the user folder in the subdirectory: .qgis2\ processing\scripts. After closing the editor window, the script appears in the toolbox (Figure 3.10).

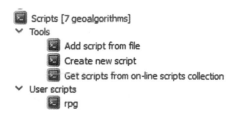

Figure 3.10. *The RPG user script into the processing toolbox*

You can run a script by double-clicking it; to edit it again, click on it with the right mouse button and choose "Edit script". In the Editor window, the button 🖎 also allows you to run the script. Since the information on the syntax of QGIS Python functions is not always easily available, an efficient way to quickly get it is to start the function from the toolbox and then to look at the file "processing.log" in the user .qgis2 \ processing directory, where successive calls of functions are logged and can be copied/pasted in your Python script.

At the beginning of the script, you can enter a list of parameters that will be requested from the user in a dialog box, preceded by a double "##". The parameter type is set to the right of the "=" sign. The parameters are then variables usable in the Python code.

```
# script Registre Parcellaire Graphique
##rpg=name
##Delimitation_ilots_RPG=vector
##Groupes_de_cultures_RPG=table
##zone_etude=vector
##zone_apprentissage=vector
##Output_Dir=folder
##Images_sat=multiple raster
```

If you run the script, you will see a dialog box for selecting input parameters (Figure 3.11).

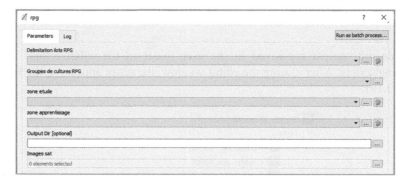

Figure 3.11. *The script interface dialog box*

The QGIS modules for Python must be specified with the "import" command.

```
#Modules Import
import sys, os
import subprocess
from qgis.core import*
import qgis.utils
import qgis.gui
from qgis.utils import iface
import processing
from PyQt4.QtCore import QFileInfo
from PyQt4.QtCore import QVariant
```

iface provides access to the QGIS interface elements. iface.mapCanvas () is the project, then fix the projection to Lambert 93 with setDestinationCrs ().

```
#Set the map Geographic system to Lambert 93

canvas = iface.mapCanvas()
canvas.mapRenderer().setProjectionsEnabled(True)
canvas.mapRenderer().setDestinationCrs(QgsCoordinateReferenceSystem("
EPSG:2154"))
```

3.3.3. *Step 1: formatting the RPG data*

Start loading the RPG data into QGIS, with "Add a vector layer" $\overset{\circ}{\mathsf{V_0}}$. The data files are: îlot_2008_076.shp, which is the ShapeFile of the limits of the FBs across the Seine-Maritime department, for the year 2008; ilot_groupe_culture_ 2008_076.csv, which is the crop composition table of the FBs; and zone_Bourville.shp, which is the limit of the catchment area of the study area (Figure 3.12).

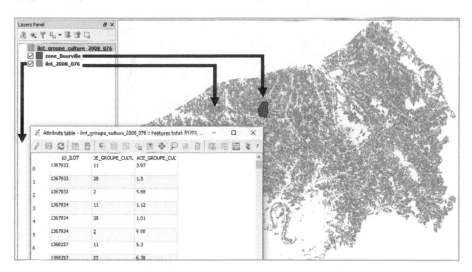

Figure 3.12. *The FB layer (ilot_2008_076) of the department of Seine-Maritime (49.65°N, 0.930°E) under QGIS. For a color version of the figure, see www.iste.co.uk/baghdadi/qgis2.zip*

First, all the layers are projected in the same projection. Indeed, even if QGIS is able to display GIS layers of different projections with the on-the-fly projection mechanism, it is advisable, for geographical operations, to project them all into the same geographic system for better accuracy. We choose here the RGF93 / Lambert 93 (EPSG: 2154) projection, which is the official French system.

* QGIS geoalgorithms ➔ Vector general tools ➔ * Reproject layer

In order to be sure to find pure FBs representing all the crops present in the Bourville catchment area, a learning zone is defined that is wider than Bourville and centered on it, being careful that the area exists in each of the images (Figure 3.13).

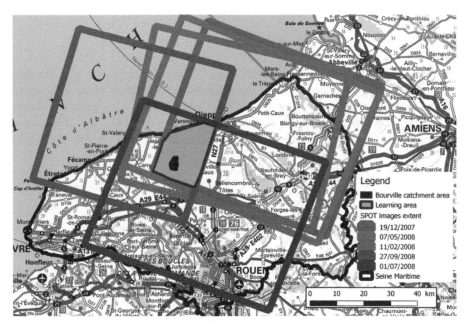

Figure 3.13. *The learning region around the Bourville catchment area (49.774°N, 0.818°E), and the five SPOT images extent, background IGN-SCAN1000 (see: www.ign.fr). For a color version of the figure, see www.iste.co.uk/baghdadi/qgis2.zip*

The FB that are in the learning region are then isolated (Figure 3.14).

* QGIS geoalgorithms ➔ Vector selection tools ➔ * Extract by location

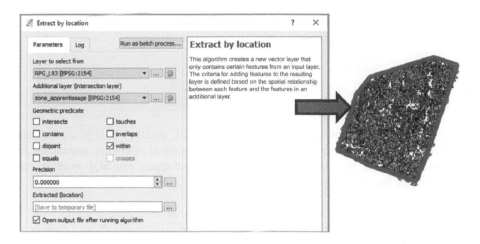

Figure 3.14. *Extraction by location of the FB located in the learning area. For a color version of the figure, see www.iste.co.uk/baghdadi/qgis2.zip*

In this way, all FBs that are inside the learning area are extracted. In the Python script, these operations are as follows:

```
# Projection of the vector layer containing the limits of the study area

zone_L93 = Output_Dir+ "/" + "zone_L93.shp"
processing.runalg("qgis:reprojectlayer",zone_etude,"EPSG:2154",
zone_L93)
# Projection of the vector layer containing the limits of the learning area
processing.runalg("qgis:reprojectlayer",zone_apprentissage,"EPSG:2154",O
utput_Dir+"/zone_Buf.shp")
zone_L93_lyr = QgsVectorLayer(zone_L93,"zone_L93", "ogr")
QgsMapLayerRegistry.instance().addMapLayer(zone_L93_lyr)
#PROJECTION OF THE RPG LAYER
RPG_L93 = Output_Dir+ "/" + "RPG_L93.shp"
out = processing.runalg("qgis:reprojectlayer",Delimitation_ilots_RPG,
"EPSG:2154",None)
# Set the ID_ILOT field in integer format
les_champs=[{'expression': u'ID_ILOT', 'length': 9, 'type': 2, 'name':
u'ID_ILOT', 'precision': 0}]
processing.runalg("qgis:refactorfields",out["OUTPUT"], les_champs
,RPG_L93)
RPG_L93_lyr = QgsVectorLayer(RPG_L93,"RPG_L93", "ogr")
QgsMapLayerRegistry.instance().addMapLayer(RPG_L93_lyr)
```

```
#EXTRACTION OF RPG FB THAT ARE INSIDE THE AREA
progress.setInfo("extraction du RPG dans l'emprise de la zone")

RPG_Zone = Output_Dir+ "/" + "RPG_zone_L93.shp"
processing.runalg("qgis:extractbylocation", RPG_L93,
Output_Dir+"/zone_Buf.shp", u'within', 0.0,RPG_Zone)
```

For the crop groups processing (CSV), we use SQL queries. As is often the case in QGIS, the same operation can be done through several operators, especially in different extensions, with different capacities in some cases. For SQL queries, the main tools are the following:

🔧 QGIS geoalgorithms ➜ Vector general tools ➜ 🔧 Execute SQL

🔧 GDAL/OGR ➜ [OGR] Miscellaneous ➜ 🔧 Execute SQL

For simple queries based only on one attribute, we can also use:

🔧 QGIS geoalgorithms ➜ Vector selection tools ➜ 🔧 Select by attribute

or:

🔧 QGIS geoalgorithms ➜ Vector selection tools ➜ 🔧 Extract by attribute

"Extract" exports the query result into a new layer, while "Select" selects the rows of the layer matching the criterion. A drawback of the "Execute SQL" is that it only works on tables associated with geometry, not on a CSV table as in our case. This limitation could be removed by creating a layer of points from the CSV by:

🔧 QGIS geoalgorithms ➜ Vector creation tools ➜ 🔧 Points layer from table

We will instead use here "Execute SQL" from GDAL/OGR, which can work directly on CSV. This first query is used to create a version of the table, sorted by FB and by area, giving the following Python code:

```
nom_fichier = os.path.basename(Groupes_de_cultures_RPG)
nom_table, extension = os.path.splitext(nom_fichier)

progress.setInfo("loop layout of each crop group (10 main groups)")
progress.setInfo(time.strftime("%H:%M:%S", time.localtime()))
progress.setPercentage(2.5)
```

```
# Convert csv to dbf, sort and conversion of area and code_groupe fields
from text to numerical
SQL = "SELECT CAST(ID_ILOT AS int) AS ID_ILOT,
CAST(CODE_GROUPE_CULTURE AS int) AS CODEG,
CAST(SURFACE_GROUPE_CULTURE AS float) AS SURFG FROM " +
nom_table + " ORDER BY ID_ILOT,SURFACE_GROUPE_CULTURE DESC"
processing.runalg("gdalogr:executesql",Groupes_de_cultures_RPG,SQL,2,O
utput_Dir+"/temp1")
```

Note that the ORDER BY directive of the SELECT command is used to sort the table in descending order (DESC), while the CAST directives are used to convert the type of some fields, as some numeric fields were imported as text from the CSV. We will then create a field (IDC) as an identifier for each crop of each FB as well as an index table for this field that considerably accelerates, sorts and joins on large size tables.

The field calculator is used for IDC calculus:

🖉 QGIS geoalgorithms ➔ Vector table tools ➔ 🖉 Field calculator

```
# Put the row number into IDC – identifier for each crop
# temporary Layer creation from table to be able to use Field calculator
out1 = processing.runalg("qgis:pointslayerfromtable",
Output_Dir+"/temp1.dbf","CODEG","CODEG","EPSG:2154",None)
processing.runalg("qgis:fieldcalculator",out1["OUTPUT"],"IDC",1,9,0,True,"$
rownum",Output_Dir+"/trie1.shp")

# indexation of IDC field => TO DO otherwise joins are slow
commande=PATH_GDAL_OGR+"ogrinfo " + Output_Dir + "/trie1.shp -sql
\"CREATE INDEX ON trie1 USING IDC\""
p = subprocess.call(commande, shell=False, startupinfo=info)
```

The "subprocess.call" Python command is used to call an external program from Python + QGIS. Here, the orginfo.exe command is used for indexing. Several times, external OGR commands will be used because some options are missing from the OGR tools built into the QGIS toolbox 🖧 GDAL/OGR . In the results table "tri1", we see the data sorted by FB (ID_ILOT) and by decreasing surface order (SURFG) (Figure 3.15).

ID_ILOT	CODEG	SURFG	IDC
483838	18	1.350000000000000	1
483840	18	1.120000000000000	2
483841	18	1.630000000000000	3
599878	24	5.030000000000000	4
599948	5	9.449999999999999	5
599948	8	7.000000000000000	6
599948	12	4.040000000000000	7
599948	24	1.700000000000000	8
600021	2	4.550000000000000	9

Figure 3.15. *Sorted crop group table*

A first Python loop will iterate over this sorted data to create a table by crop group (CODEG): G1.shp, G2.shp, G3.shp, …, Gn.shp.

```
max_grp = 10

for grp in range(1,max_grp+1):
    groupe="G"+str(grp)
    progress.setInfo(str(grp)+" / 10")
    table_tri="trie"+str(grp)
    # isole le groupe
    commande = PATH_GDAL_OGR+"ogr2ogr -overwrite -dialect sqlite -sql
\"SELECT CAST(ID_ILOT AS int) AS ID_ILOT2, CAST(CODEG AS int) AS
CODEG2, CAST(SURFG AS float) AS SURFG2, CAST(IDC AS int) AS IDC2 FROM
" + table_tri + " GROUP BY ID_ILOT HAVING MIN(IDC) ORDER BY IDC\" " +
Output_Dir+"/temp_"+groupe + ".dbf " + Output_Dir+"/"+table_tri+".dbf"
    p = subprocess.call(commande, shell=False, startupinfo=info)

... see script listing for more details
```

The external command ogr2ogr.exe is here used to make the SQL query, because it accounts for the option "GROUP BY ID_ILOT HAVING MIN (IDC) ORDER BY IDC", which is not supported by "Execute SQL" and which allows us to identify the information group by group throughout iterations. An SQL function is used to count the crops in each FB, and then to join them to the FB.

```
# statistics by category to count the crops in each FB
stat_cat = Output_Dir+ "/" + "stat_cat.dbf"

out1 =
processing.runalg("qgis:pointslayerfromtable",Groupes_de_cultures_RPG,"I
D_ILOT","ID_ILOT","EPSG:2154",None)
processing.runalg("qgis:fieldcalculator",out1["OUTPUT"],"SURF",0,10,2,
True,"SURFACE_GR",Output_Dir+"/grp_cultures")

# SQL with ogr2ogr much quicker than qgis:statisticsbycategories
#processing.runalg('qgis:statisticsbycategories',
out2["OUTPUT_LAYER"] ,'SURF','ID_ILOT',stat_cat)

commande = PATH_GDAL_OGR+"ogr2ogr -overwrite -dialect sqlite -sql
\"SELECT CAST(ID_ILOT AS int) AS ID_ILOT, count(ID_ILOT) as NBG,
sum(SURF) as SURF_ILOT FROM grp_cultures GROUP BY ID_ILOT\" " +
Output_Dir+"/stat_cat.dbf " + Output_Dir+"/grp_cultures.dbf"
p = subprocess.call(commande, shell=False, startupinfo=info)

#join statistics to the RPG
progress.setInfo("stats join on RPG")
progress.setInfo(time.strftime("%H:%M:%S", time.localtime()))
progress.setPercentage(14.5)
Join_RPG_Zone = Output_Dir+ "/" + "ilots_L93.shp"
out1 = processing.runalg('qgis:joinattributestable', RPG_Zone, stat_cat,
'ID_ILOT', 'ID_ILOT', None)
```

At the end of this step, the NBG field gives the number of crop groups, and SURF_ILOT, the total area of the crops within the FB. A second Python loop on the crop groups will be used to integrate the Gn.shp files produced by the first loop by joining them in the FB:

```
# join groups data
progress.setInfo("import of each group data into FBs")
progress.setInfo(time.strftime("%H:%M:%S", time.localtime()))
for grp in range(1,max_grp+1):
    groupe="G"+str(grp)
    progress.setInfo(str(grp)+" / 10")
    out_temp[grp+2] = processing.runalg('qgis:joinattributestable',
out_temp[grp+1]["OUTPUT_LAYER"], Output_Dir+"/"+groupe+".dbf",
'ID_ILOT', 'ID_ILOT', None)
```

```
processing.runalg("qgis:saveselectedfeatures",out_temp[max_grp+2]["OUT
PUT_LAYER"], Join_RPG_Zone)

join_lyr = QgsVectorLayer(Join_RPG_Zone,"join_L93", "ogr")
QgsMapLayerRegistry.instance().addMapLayer(join_lyr)
```

At the end, the "ilots_L93.shp" layer of the FB (ilot) is obtained with all the groups' information in columns (Figure 3.16). For each crop c, codeg_c is the code of the crop group and surfg_c, the area of this group.

	ID_ILOT	NBG	sum	SURF_ILOT	SURF_POLY	codeg_2	idc_2	surfg_2
2583	4310428	10	98.05	98.05	97.94	5	58879	4.560000000000...
1760	4297309	10	66.62	66.62	66.55	24	38312	8.100000000000...
1295	4290368	7	88.04	88.04	87.94	5	27786	7.550000000000...
3532	4323884	5	48.11	48.11	48.06	5	79798	6.700000000000...
3059	4316809	5	40.91	40.91	40.86	24	68836	6.000000000000...

Table attributaire - ilots_L93 :: Total des entités: 3763, filtrées: 3763, sélectionnées: 0 {1 ?} {2,?} {3

Figure 3.16. *Layer ilots_L93 table, which now contains all the information of the crop groups as fields: codeg_x is the code and surfg_x the area, for the xth crop*

A 10 m reduced size version of the FB is created, in order to limit edge effects during subsequent statistical analysis of the SPOT images in FB. To do this, we use the "buffer distance" function of QGIS:

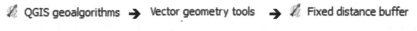 QGIS geoalgorithms ➔ Vector geometry tools ➔ Fixed distance buffer

```
# -10m buffer area of FB to prevent edge effects
progress.setInfo("buffer zones of FBs")

processing.runalg("qgis:fixeddistancebuffer",Output_Dir+"/ilots_L93.shp",-
10,5,False,Output_Dir+"/temp_buf.shp")
# check as buffer generates non valid entities
processing.runalg("qgis:checkvalidity",Output_Dir+"/temp_buf.shp",0,Outp
ut_Dir+"/temp_buf_ok.shp",None,None)
# SURF_BUF is the buffer area
processing.runalg("qgis:fieldcalculator",Output_Dir+"/temp_buf_ok.shp","S
URF_BUF",0,10,2,True,"$area /10000.0",
Output_Dir+"/ilots_L93_buf.shp")
```

Note the negative value of the distance (−10) in the buffer function, to obtain a smaller area than the original FB (Figure 3.17)

Figure 3.17. *A buffer area in red on an FB parcel in green. For a color version of the figure, see www.iste.co.uk/baghdadi/qgis2.zip*

From the ilots_L93_buf.shp layer, we finally export the pure FBs (only one crop NBG = 1) and the poly FBs (NBG > 1) by a request:

🖋 QGIS geoalgorithms ➜ Vector selection tools ➜ 🖋 Extract by attribute

```
# Export pure FB = only one crop
processing.runalg("qgis:extractbyattribute",Output_Dir+"/ilots_L93_buf.shp
","NBG",0,"1",Output_Dir+"/ilots_purs_buf.shp")
# Export poly FB nb > 1 crop
processing.runalg("qgis:extractbyattribute",Output_Dir+"/ilots_L93_buf.shp
","NBG",2,"1",Output_Dir+"/ilots_poly_buf0.shp")
processing.runalg("qgis:extractbylocation",Output_Dir+"/ilots_poly_buf0.sh
p", Output_Dir+"/zone_L93.shp", u'within',
0.0,Output_Dir+"/ilots_poly_buf.shp")
```

3.3.4. *Step 2: classification of SPOT satellite Images*

For image classification, before starting the script, you must have loaded the images into QGIS with "Add a raster layer". At the beginning of the script, a multiple-choice dialog box is used to select the images to be processed (Figure 3.18).

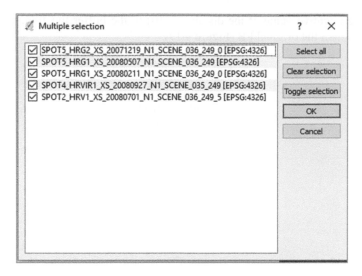

Figure 3.18. *Multiple choice dialog box for satellite images*

The Python code line for creating this dialog box is as follows:

```
##Images_sat=multiple raster
```

The Images_sat variable contains the user choice, in the form of a string where the different images are separated by a ";" character. A Python loop will iterate over the SPOT images to apply the following pre-processing operations:

– projection of images in Lambert 93 geographic system:

– masking of images by the contours of the learning area:

– NDVI calculation with OTB:

The Python code for these steps is as follows:

```
lesNDVI=""

n = 1
for im in lesImages:
    progress.setInfo(im)
    progress.setInfo(time.strftime("%H:%M:%S", time.localtime()))
    fileInfo = QFileInfo(im)
    baseName = fileInfo.baseName()
    chemin = fileInfo.absolutePath()
    print(baseName)
    spot_lyr = QgsRasterLayer(im, baseName)
    #print(spot_lyr)
    proj = spot_lyr.crs().authid()
    spot_L93 = Output_Dir+"/"+baseName+"_L93.tif"
    processing.runalg("gdalogr:warpreproject", im, "", "EPSG:2154", "",
10,0,5,0,75,6,1,False,0,False,"",spot_L93)
    # decoupage de la zone
    decoupe = Output_Dir+"/"+baseName+"_mask_L93.tif"

processing.runalg("gdalogr:cliprasterbymasklayer",spot_L93,zone_etude,
"",False,True,False,5,0,75,6,1,False,0,False,"",decoupe)

    # NDVI computation
    ndvi = Output_Dir+"/"+baseName+"_NDVI_L93.tif"

processing.runalg("otb:radiometricIndices",decoupe,128,0,1,2,3,1,0,ndvi)
    lesNDVI = lesNDVI+ndvi+";"
    progress.setPercentage(17+n*3)
    n = n + 1
```

At the end of the Python loop, a NDVI channel is available for each of the initial SPOT images. These channels are grouped in the same composite image, with a merge operation (Figure 3.19).

GDAL/OGR ➜ [GDAL] Miscellaneous ➜ Merge

```
# NDVI are merged in the same composite image
progress.setInfo("regroupement des NDVI dans une image composite")
progress.setInfo(time.strftime("%H:%M:%S", time.localtime()))
compoNDVI = Output_Dir+"/"+"compoNDVI.tif"
```

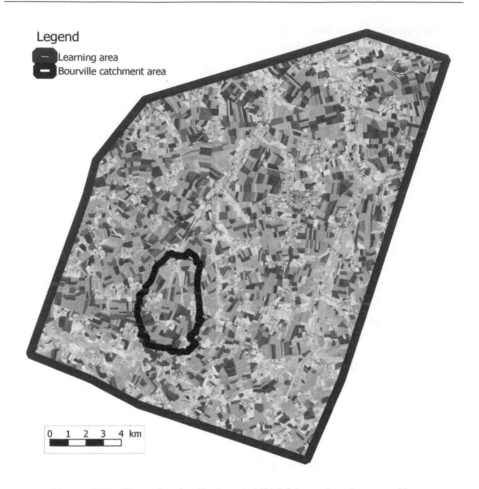

Figure 3.19. *Example of a diachronic NDVI false colored composition (49.774°N, 0.818°E): February 11, 2008 (red), May 7, 2008 (green) and September 27, 2008 (blue) inside the learning area. For a color version of the figure, see www.iste.co.uk/baghdadi/qgis2.zip*

Half of the pure FB extracted in step 1 are used as learning zones or ROI to characterize each crop group by its NDVI statistics at the different dates. These statistics are then used to classify the NDVI image into crop groups. It was found that a number of FBs, which were defined as pure in the RPG, were actually not pure in the images (Figure 3.20). In order to select the truly pure islands, a first Bayes classification is made:

Orfeo Toolbox → Learning → TrainImagesClassifier (bayes)

 Orfeo Toolbox → Learning → Image Classification

The "CODEG" field contains the crop group code.

```
# First classification to test pure FB and just keep really pure ones
# Learning stat Bayes
progress.setInfo("classif recherche ilot purete > 70%")
progress.setInfo(time.strftime("%H:%M:%S", time.localtime()))

processing.runalg("otb:trainimagesclassifierbayes",compoNDVI,
Output_Dir+"/ilots_purs_buf.shp","",0,1000,1000,1,True,0.5,"CODEG",0,0,
Output_Dir+"/matrix.xml",Output_Dir+"/model.xml")

# classif Bayes
processing.runalg("otb:imageclassification",compoNDVI,None,
Output_Dir+"/model.xml","",128,Output_Dir+"/bayes_purs.tif")
```

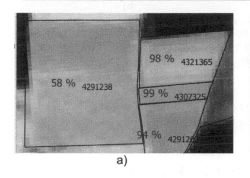
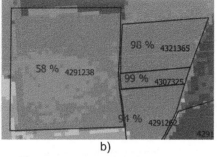

a)   b)

Figure 3.20. *FB defined as "pure" in the RPG: a) display on the NDVI colored composition and b) on the Bayes classification. FB 4291238, which appears heterogeneous, will not be retained as a pure parcel. For a color version of the figure, see www.iste.co.uk/baghdadi/qgis2.zip*

From the result of this first Bayes classification, we look at the most frequent class in each FB and its area. Only those parcels for which this area represents more than 70% of the total FB area should be retained as really pure. The resulting layer is then called "ilots_purs70.shp". For this operation, the steps are:

– rasterization of pure FB:

    GRASS GIS 7 commands → Vector (v.*) → v.to.rast.attribute

– calculation of the class statistics (modal value) by FB:

🌿 GRASS GIS 7 commands ➜ Raster (r.*) ➜ 🌿 r.statistics

– selection of the pixels within the FB, corresponding to the modal class (Figure 3.21):

🌍 SAGA (2.1.2) ➜ Raster calculus ➜ 🌍 Raster calculator

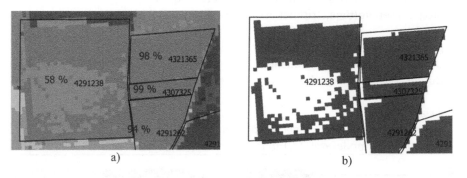

a)   b)

Figure 3.21. *FB defined as "pure" in the RPG: a) display on the Bayes classification and b) display on the class corresponding to the majority surface. For a color version of the figure, see www.iste.co.uk/baghdadi/qgis2.zip*

– counting the number of pixels corresponding to the modal class; for this we calculate the sum of the preceding pixels that are binary (1/0), and the result is a raster layer:

🌿 GRASS GIS 7 commands ➜ Raster (r.*) ➜ 🌿 r.statistics

– vectorization of the previous layer:

🌿 GRASS GIS 7 commands ➜ Raster (r.*) ➜ 🌿 r.to.vect

– extraction of the polygons' centroids of the vector layer.

In the vector geometry tools, there is a function for calculating centroids:

🔧 QGIS geoalgorithms ➜ Vector geometry tools ➜ 🔧 Polygon centroids

However, as Figure 3.22 shows, the centroid being the "center of gravity" of the polygon, the point is not necessarily within the polygon. Since we want to use this point as a marker of the polygon, we will rather use the function "random points inside polygons", which generates a point at a random position, but which is located inside the polygon.

📁 QGIS geoalgorithms ➜ Vector creation tools ➜ 📁 Random points inside polygons (fixed)

One point per polygon is generated, with a minimum distance of 5 m from the edge of the polygon.

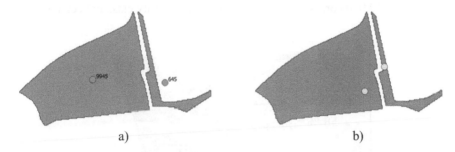

a)   b)

Figure 3.22. *Extraction of centroids using: a) the "centroids of polygons" function and b) the "random points inside polygons" function.*

The step of point creation has "lost" the information fields of the vectorized polygons, in particular the "value" field that contains the number of pixels of the modal class. It can be retrieved by a spatial join of the polygons to the points, with the relation "is inside":

📁 QGIS geoalgorithms ➜ Vector general tools ➜ 📁 Join attributes by location

Spatial join, or geographic join, allows two layers of information to be related not by using table fields, as is the case for an attribute join, but by using a geographic relationship: the point is related to the FB in which it is included. A spatial join of the centroids toward the FBs allows the "value" field of the FB to be recovered.

Calculation of the "purity" percentage of the FB, as a function of the modal class surface, uses the field calculator:

📁 QGIS geoalgorithms ➜ Vector table tools ➜ 📁 Field calculator

By calculation, a new "pure" field is created, which gives the area ratio of the FB covered by the majority class (Figure 3.23):

$$pure = \frac{value \times 25}{10,000 \times AREA\_BUF}$$ [3.3]

where value is the number of pixels of the modal class and AREA_BUF is the area in hectares of the FB (buffer version), the resolution of the previous raster being 5 m; value × 25/10,000 corresponds to the area of the majority class in hectares.

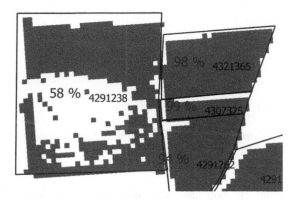

Figure 3.23. *Percentage of the FB area, covered by the majority class (in red). For a color version of the figure, see www.iste.co.uk/baghdadi/qgis2.zip*

The extraction of the FBs with "purity" greater than 70% is done by a simple extraction by attribute operation:

🖉 QGIS geoalgorithms ➜ Vector selection tools ➜ 🖉 Extract by attribute

At the end of this step, with "Extract by Attribute" (pure > 0.7), we create the layer "ilots_purs70.shp" of the new pure islands, which we use as "ROI" for the following classifications.

Three image classification methods have been used, each with a step of acquiring the reference statistics on the previous ROIs, followed by the actual classification step of the image from these references:

– Bayes classification method:

🗿 Orfeo Toolbox ➜ Learning ➜ 🗿 TrainImagesClassifier (bayes)

 Orfeo Toolbox ➔ Learning ➔  Image Classification

– SVM classification method:

Orfeo Toolbox ➔ Learning ➔ TrainImagesClassifier (libsvm)

Orfeo Toolbox ➔ Learning ➔ Image Classification

– Random Forest classification method:

Orfeo Toolbox ➔ Learning ➔ TrainImagesClassifier (rf)

Orfeo Toolbox ➔ Learning ➔ Image Classification

For the three classifications, several post-processing operations are applied:

– masking by the "buffer" FB layer:

GDAL/OGR ➔ [GDAL] Extraction ➔ Clip raster by mask layer

majority filtering, which makes it possible to "smooth" the result by removing isolated pixels:

SAGA (2.1.2) ➔ Raster filter ➔ Majority filter

– growing regions, which makes it possible to return to the initial size of the FB that had been reduced by the buffer effect:

GDAL/OGR ➔ [GDAL] Analysis ➔ Fill nodata

– calculation of the confusion matrix with the pure "validation" FB:

Orfeo Toolbox ➔ Learning ➔ ComputeConfusionMatrix (vector)

The following Python code corresponds to the case of the Bayes classification:

```
# Bayes classif with ilots_purs70 as training area
progress.setInfo("classification de Bayes")
progress.setInfo(time.strftime("%H:%M:%S", time.localtime()))
```

```
processing.runalg("otb:trainimagesclassifierbayes",compoNDVI,
Output_Dir+"/ilots_purs70.shp","",0,1000,1000,1,True,0.5,"CODEG",0,0,
Output_Dir+"/matrixBayes.xml",Output_Dir+"/modelBayes.xml")
processing.runalg("otb:imageclassification",compoNDVI,None,
Output_Dir+"/modelBayes.xml","",128,
Output_Dir+"/clas_bayes_brut.tif")
# postprocessing
#masking  to clean what is outside of the parcels and edge effects
out=processing.runalg("gdalogr:cliprasterbymasklayer",
Output_Dir+"/clas_bayes_brut.tif",Output_Dir+"/ilots_L93_buf.shp","0",
False,False,False,0,0,75,6,1,False,0,False,"",None)
#Majority  filter 3x3
processing.runalg("saga:majorityfilter",out["OUTPUT"],1,1,0,
Output_Dir+"/temp_otb1.tif")
# pseudo growing region to get back to the initial size of parcel

processing.runalg("gdalogr:fillnodata",Output_Dir+"/temp_otb1.tif",2,10,1,
None,False,Output_Dir+"/clas_bayes.tif")
# calculation of the confusion matrix
processing.runalg("otb:computeconfusionmatrixvector",
Output_Dir+"/clas_bayes.tif",0,Output_Dir+"/ilots_purs70_valide.shp",
"CODEG",0,128,Output_Dir+"/matrix2Bayes.csv")
```

Merge the three classifications by the Dempster–Shafer method to create the final classification:

Orfeo Toolbox → Learning → FusionOfClassifications (dempstershafer)

In the case where, for the same pixel, two classifications give two different classes but with a close probability, the merge step does not classify the pixel and assigns it to "NoData" value, which results in holes in the final map. A region growth operation (five iterations) is thus used to fill these holes (Figure 3.24):

SAGA (2.1.2) → Raster tools → Shrink and expand

a)   b)

Figure 3.24. *Geometric post-processing of the classification*
a) by growing region and b) in order to fill the holes in the map. For a
color version of the figure, see www.iste.co.uk/baghdadi/qgis2.zip

Using the pure FB polygons that were not used during the learning phase (50% of the FBs), the confusion matrix is calculated to evaluate the result quality:

Orfeo Toolbox ➔ Learning ➔ ComputeConfusionMatrix (vector)

The corresponding Python code is as follows:

```
#Merge the 3 classifications dempster-shafer method
progress.setInfo("Fusion des 3 classifications: METHODE DE DEMPSTER-
SHAFER")
progress.setInfo(time.strftime("%H:%M:%S", time.localtime()))

processing.runalg("otb:fusionofclassificationsdempstershafer",
Output_Dir+"/clas_bayes.tif;"+Output_Dir+"/clas_svm.tif;"+Output_Dir+
"/clas_RF.tif",0,Output_Dir+"/matrix2Bayes.csv;"+Output_Dir+
"/matrix2svm.csv;"+Output_Dir+"/matrix2RF.csv",0,0,0,Output_Dir+
"/clas_Dempster_shafer_brut.tif")

# growing region 5 X
progress.setInfo("post-traitement de la classification")
progress.setInfo(time.strftime("%H:%M:%S", time.localtime()))

processing.runalg("saga:majorityfilter",Output_Dir+
"/clas_Dempster_shafer_brut.tif",1,3,0,Output_Dir+"/tempsh1.tif")
processing.runalg("saga:shrinkandexpand",Output_Dir+
"/tempsh1.tif",1,0,1,3,Output_Dir+"/tempsh2.tif")
processing.runalg("saga:shrinkandexpand",Output_Dir+
"/tempsh2.tif",1,0,1,3,Output_Dir+"/tempsh3.tif")
processing.runalg("saga:shrinkandexpand",Output_Dir+
"/tempsh3.tif",1,0,1,3,Output_Dir+"/tempsh4.tif")
processing.runalg("saga:shrinkandexpand",Output_Dir+
"/tempsh4.tif",1,0,1,3,Output_Dir+"/tempsh5.tif")
```

```
processing.runalg("saga:shrinkandexpand",Output_Dir+
"/tempsh5.tif",1,0,1,3,Output_Dir+"/clas_Dempster_shafer.tif")

# masking with poly FB
processing.runalg("grass7:v.to.rast.value",Output_Dir+"/ilots_poly.shp",
0,1,etendue,10,-1,0.0001,Output_Dir+"mask_ilot_poly.tif")
processing.runalg("saga:gridmasking",Output_Dir+
"/clas_Dempster_shafer.tif",Output_Dir+"mask_ilot_poly.tif",Output_Dir+"/
clas_Dempster_shafer_poly0.tif")
processing.runalg("saga:majorityfilter",Output_Dir+
"/clas_Dempster_shafer_poly0.tif",1,3,0,Output_Dir+
"/clas_Dempster_shafer_poly.tif")

# confusion matrix
progress.setInfo("Calcul de la matrice de confusion")
progress.setInfo(time.strftime("%H:%M:%S", time.localtime()))

processing.runalg("otb:computeconfusionmatrixvector",Output_Dir+
"/clas_Dempster_shafer.tif",0,Output_Dir+"/ilots_purs70_valide.shp",
"CODEG",0,128,Output_Dir+"/matrix2Dempster_shafer.csv")
```

The final result of the classification steps is shown in Figure 3.25.

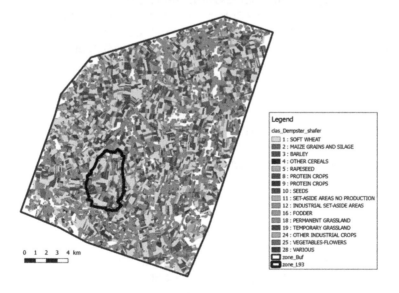

Figure 3.25. *Final classification result. For a color version of the figure, see www.iste.co.uk/baghdadi/qgis2.zip*

To estimate the quality of this classification, we analyze the confusion matrix that was produced during the merging operation according to the Dempster–Shafer method (Table 3.2). The user precision (%u) is the percentage of pixels of a crop group in the classified image, which corresponds to the same group in the validation parcels. For groups: soft wheat (1), barley (3), rapeseed (5) and permanent grassland (18), the user accuracy is greater than 90%. It is 64% for corn (2) and 85% for vegetable-flowers (25). For some groups, such as set-aside (11), we cannot conclude because numbers are too small.

		CLASSIFICATION															
		1	2	3	5	8	9	10	11	16	18	19	24	25	28	total	%p
REFERENCE	1	98945	17	1619	0	0	0	30	0	0	1132	104	0	0	13	101860	97.1
	2	10	7392	23	0	7	5	14	0	709	181	134	1513	304	0	10292	71.8
	3	1918	105	17488	2	0	190	159	0	0	144	0	0	0	0	20006	87.4
	5	191	0	0	2901	0	0	0	0	0	860	0	0	0	0	3952	73.4
	8	0	17	0	0	348	0	0	0	0	0	0	0	0	0	365	95.3
	9	0	45	1	101	0	607	2524	0	0	0	0	0	0	0	3278	18.5
	10	1	638	0	0	14	2151	10700	0	0	30	0	0	0	7	13541	79.0
	11	4	0	0	0	0	0	0	0	0	691	0	0	0	23	718	0.0
	16	0	0	0	0	0	0	0	0	0	0	0	104	0	0	104	0.0
	18	277	40	239	0	0	0	5	85	0	88691	1751	0	0	1451	92539	95.8
	19	0	2	14	0	0	0	0	89	0	2126	310	0	0	0	2541	12.2
	24	3	882	2	2	7	10	7	0	242	5	0	4952	854	5	6971	71.0
	25	65	2370	1	144	603	200	182	0	6	17	0	2304	11410	0	17302	65.9
	28	0	0	0	0	4	5	0	0	0	2	0	2	808	0	821	0.0
	total	101415	11510	19390	3155	991	3177	13631	185	973	93897	2318	8899	13401	1527	274290	
	%u	97.6	64.2	90.2	91.9	35.1	19.1	78.5	0.0	0.0	94.5	13.4	55.6	85.1	0.0		

Table 3.2. *Confusion matrix of the classification, calculated on 50% of the pure parcels, which have not been used for learning. %p is producer accuracy in percent and %u is user precision*

We can compare the classification to the parcels map resulting from the 2008 field survey (Figure 3.26): the parcels' geometry is well recognized. There are, however, two types of problem:

–geometric, due to irregular spots in some plots, often due to crop heterogeneity, which is too confusing for the classification;

typological, for which plot geometry proves to be exact but is not assigned to the same class as the reference plots. This is the case for field plots in "7: other oilseeds" (Figure 3.26(b)) and classified as "10: seeds" (Figure 3.26(a)). It is a problem inherent to the RPG nomenclature, according to which some crops can potentially be assigned to different RPG stations, depending on their use and not their nature: seed, set-aside, industrial set-aside, etc.

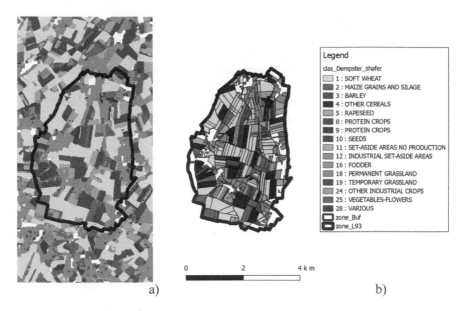

Legend

das_Dempster_shafer
- 1 : SOFT WHEAT
- 2 : MAIZE GRAINS AND SILAGE
- 3 : BARLEY
- 4 : OTHER CEREALS
- 5 : RAPESEED
- 8 : PROTEIN CROPS
- 9 : PROTEIN CROPS
- 10 : SEEDS
- 11 : SET-ASIDE AREAS NO PRODUCTION
- 12 : INDUSTRIAL SET-ASIDE AREAS
- 16 : FODDER
- 18 : PERMANENT GRASSLAND
- 19 : TEMPORARY GRASSLAND
- 24 : OTHER INDUSTRIAL CROPS
- 25 : VEGETABLES-FLOWERS
- 28 : VARIOUS
- zone_Buf
- zone_L93

0   2   4 k m

a)   b)

Figure 3.26. *Comparison of the classification, a) with the plot map and
b) "field truth" of the same crop year. For a color version of
the figure, see www.iste.co.uk/baghdadi/qgis2.zip*

3.3.5. *Step 3: intersect overlay between extracted AP and FB and crop validation*

This last step aims to remove the uncertainties outlined at the end of step 2. The classification results are compared, FB by FB, with the RPG information of the poly FB. To do this, it is necessary to have "objects" plots, derived from the classification that is a raster (tif image). By definition, in a raster, we only have pixel-to-pixel information; there is no object. It is therefore necessary to transform the result of the classification into objects by a vectorization:

Commandes GRASS GIS 7 → Raster (r.*) → r.to.vect

Holes in the vector layer are filled:

QGIS geoalgorithms → Vector geometry tools → Fill holes

Vectors created by vectorization of a raster often have "staircase" limits, due to the fact that the algorithm follows the contours of the square pixels. The simplification of these contours can then be done by vector generalization (Figure 3.27).

🌱 GRASS GIS 7 commands ➜ Raster (r.*) ➜ 🌱 v.generalize

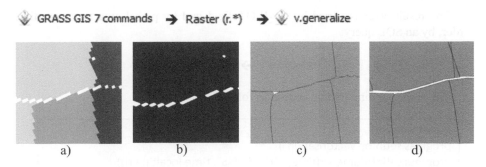

a)   b)   c)   d)

Figure 3.27. *Vectorization steps: a) the classified image, b) the vectorized layer, c) the generalized one and d) the overlay intersect with the boundaries of the FB. For a color version of the figure, see www.iste.co.uk/baghdadi/qgis2.zip*

In order to integrate the RPG information of the FB that contains the classified parcel, an overlay intersection operation is performed between the vectorized plot and the poly FB:

🖉 QGIS geoalgorithms ➜ Vector overlay tools ➜ 🖲 Intersection

Then, for each plot, its area and an index to characterize its shape are calculated:

🖉 QGIS geoalgorithms ➜ Vector table tools ➜ 🖉 Field calculator

$area variable in a field calculator formula gives the area and $perimeter the perimeter. ShapeIndex = $perimeter/(2 * sqrt (3.14159 * $area)) is computed as the shape index of the polygons. To eliminate the last artifacts, only polygons with a surface area > 0.01 ha are retained:

🖉 QGIS geoalgorithms ➜ Vector selection tools ➜ 🖉 Extract by attribute

Each parcel is now identified by its FB identifier following the overlay intersection, the number of plots per FB that can be calculated. The count is done by a statistic per category according to the identifier field "ID_LOT". The statistical table created is then joined in the table of the parcels, to recover the "count" field:

🖉 QGIS geoalgorithms ➜ Vector table tools ➜ 🖉 Statistics by categories

🖉 QGIS geoalgorithms ➜ Vector general tools ➜ 🖉 Join attributes table

The result layer is sorted by ID_ILOT field and by parcel area in descending order, by an SQL query:

 GDAL/OGR ➜ [OGR] Miscellaneous ➜ Execute SQL

The corresponding Python code is as follows:

```
# vectorization
progress.setInfo("vectorisation")
progress.setInfo(time.strftime("%H:%M:%S", time.localtime()))

processing.runalg("grass7:r.to.vect",Output_Dir+
"/clas_Dempster_shafer_poly.tif",2,True,etendue,3,Output_Dir+
"/parcelles_brut.shp")
# bouche les trous
progress.setInfo("lissage generalisation")
progress.setInfo(time.strftime("%H:%M:%S", time.localtime()))

processing.runalg("qgis:fillholes",Output_Dir+"/parcelles_brut.shp",
100000,Output_Dir+"/parcelles_brut_fill.shp")
#Lissage des parcelles
processing.runalg("grass7:v.generalize",Output_Dir+
"/parcelles_brut_fill.shp",11,50,7,50,0.5,3,0,0,0,1,1,10,False,True,
etendue,-1,0.0001,0,Output_Dir+"/parcelles_lisses.shp")
#croisement avec les ilots
progress.setInfo("croisement avec les ilots RPG")
progress.setInfo(time.strftime("%H:%M:%S", time.localtime()))

processing.runalg("saga:intersect",Output_Dir+"/parcelles_lisses.shp",
Output_Dir+"/ilots_poly.shp",True,Output_Dir+"/parcelles_inter.shp")
# indice de forme: ShapeIndex (Perimeter / (2 * SquareRoot(PI * Area))
processing.runalg("qgis:fieldcalculator",Output_Dir+"/parcelles_inter.shp","
ShapeIndex",0,10,2,True,
" $perimeter / (2 * sqrt(3.14159 * $area))",
Output_Dir+"/parcelles_indices.shp")
# calcul de la surface des parcelles
out = processing.runalg("qgis:fieldcalculator",Output_Dir+
"/parcelles_indices.shp","surface",0,10,2,True," $area /10000.0",None)
# ne garde que les polygones de surface > 0.1 ha
out4 = processing.runalg("qgis:extractbyattribute",
out["OUTPUT_LAYER"],"surface",2,"0.1",None)
# count parcel number by FB
```

```
out2 =
processing.runalg("qgis:statisticsbycategories",out4["OUTPUT"],"surface","I
D_ILOT",None)
out3 = processing.runalg('qgis:joinattributestable', out4["OUTPUT"],
out2["OUTPUT"], 'ID_ILOT', 'category',None )
out5= processing.runalg("qgis:fieldcalculator",out3["OUTPUT_LAYER"],
"CODEG_P",0,10,0,True,"0",None)
out6= processing.runalg("qgis:fieldcalculator",out5["OUTPUT_LAYER"],
"SURFG_P",0,10,1,True,"0",None)
processing.runalg("qgis:fieldcalculator",out6["OUTPUT_LAYER"],
"NB_PARC",0,8,0,True," count",Output_Dir+"/parcelles_poly.shp")
# SORT by FB and by surface descending order
SQL = "SELECT  * FROM parcelles_poly ORDER BY ID_ILOT,surface DESC"
processing.runalg("gdalogr:executesql",Output_Dir+
"/parcelles_poly.shp",SQL,2,Output_Dir+"/parcelles_poly_trie.shp")
```

Figure 3.28 displays the output classification after these vectorization steps.

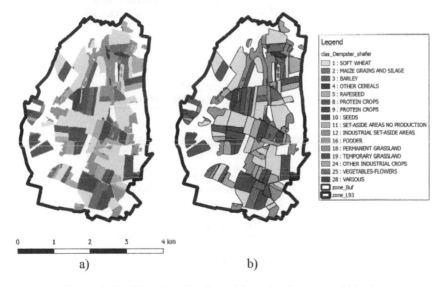

Figure 3.28. *The classification a) in raster format and b) after vectorization and integration with the poly FB. For a color version of the figure, see www.iste.co.uk/baghdadi/qgis2.zip*

In the final part of the processing, a Python loop iterates across the whole parcels table in order to manage remaining errors, parcel by parcel. The Qgs VectorLayer function gives a descriptor that makes it possible to manage a QGIS layer in Python and layer.getFeatures returns a stream of all the layer polygons. It is thus easy to iterate through all the parcels, with a simple "for" Python loop.

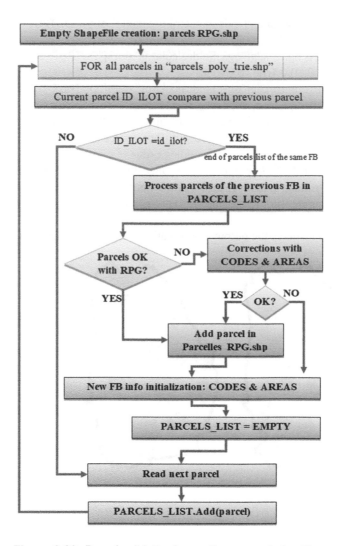

Figure 3.29. *Parcel validation/correction general algorithm*

Since the table has been previously sorted according to the "ID_ILOT" field, the parcels of the same FB therefore lie in successive lines. The general algorithm is given in Figure 3.29. The "corrections using CODES and AREAS" section is dedicated to analyzing and, whenever possible, correcting differences between the parcel information and the RPG FB information. It is still in development and has only been implemented for a limited number of simple cases. In the simplest case (Figure 3.30), the information (codes and areas) resulting from the classification directly matches the RPG information, and the parcels are then directly validated. Figure 3.30 shows that for FB 4277220, the image processing finds the same three crop classes (10, 25 and 1), and when the areas of the same crop are totalized to compare with RPG surfaces: soft wheat (1), 14.4 ha (14.44 ha RPG); flower-growing vegetables (25), 13.41 ha (13.02 ha RPG) and seeds (10), 11.79 ha (12.9 ha RPG), the resulting surfaces are very close to those of the RPG.

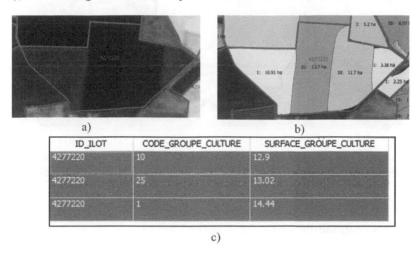

a)   b)

ID_ILOT	CODE_GROUPE_CULTURE	SURFACE_GROUPE_CULTURE
4277220	10	12.9
4277220	25	13.02
4277220	1	14.44

c)

Figure 3.30. *Final result, FB 4277220 case. a) FB display on NDVI color composition, b) resulting parcels with crop code and area, c) corresponding RPG data for this FB. For a color version of the figure, see www.iste.co.uk/baghdadi/qgis2.zip*

Among the 56 polycultural FBs that are in the small catchment area of Bourville, when only considering the crops occupying more than 0.5 ha, there are only 10 FBs for which the crops found by image processing do not exactly match the RPG crops: 15 if taking into account those less than 0.5 ha (Figure 3.31). To do this calculation, we use joins, statistics by categories and selections/extractions:

🔧 QGIS geoalgorithms ➜ Vector general tools ➜ 🔧 Join attributes table

🔧 QGIS geoalgorithms ➜ Vector table tools ➜ 🔧 Statistics by categories

🔧 QGIS geoalgorithms ➜ Vector selection tools ➜ 🔧 Extract by attribute

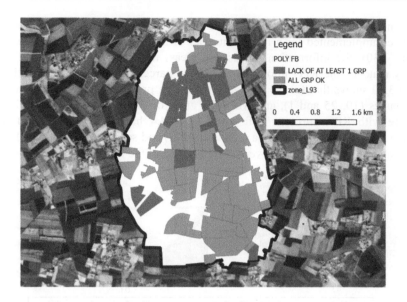

Figure 3.31. *Concordance between crops found by image processing and crops described in the poly FBs of the RPG (green: matching; red: non-matching). For a color version of the figure, see www.iste.co.uk/baghdadi/qgis2.zip*

3.4. Acknowledgments

The authors would like to thank the AFB for funding the work presented previously in the context of the PACS-AAC Action (Agricultural Pressure and Socio-Economic Coordination on AAC), as well as AREAS, which transmitted field monitoring data for the Bourville site.

3.5. Bibliography

[LEP 16] LE PRIOL C., Prototypage d'un outil de cartographie automatique des cultures agricoles à l'échelle parcellaire à partir d'images de télédétection et du Registre parcellaire graphique, Mastère spécialisé SILAT, AgroParisTech, Paris, France, 2016.

[MAR 14] MARTIN P., RONFORT C., LAROUTIS D. *et al.*, "Cost of best management practices to combat agricultural runoff and comparison with the local populations' willingness to pay: case of the Austreberthe watershed (Normandy, France)", *Land Use Policy*, vol. 38, pp. 454–466, 2014.

[MAR 17] MARTIN P., "RPG explorer – Cours en ligne", AgroParisTech Paris, France, Consulted April 19, 2017, available at https://tice.agroparistech.fr/coursenligne/courses/RPGEXPLORER/index.php, 2017.

[SAG 08] SAGRIS V., DEVOS W., LPIS Core Conceptual Model: Methodology for Feature Catalogue and Application Schema, GeoCAP Discussion Paper, European Commission, DG Joint Research Center Ispra, Institute for the Protection and Security of the Citizen, Agriculture Unit, p. 59, 2008.

[SHA 76] SHAFER G., *A Mathematical Theory of Evidence*, Princeton University Press, Princeton, 1976.

Land Cover Mapping Using Sentinel-2 Images and the Semi-Automatic Classification Plugin: A Northern Burkina Faso Case Study

4.1. Context

The African population is today estimated at 2 billion and is expected to double by 2050. The concerns for agricultural production and food security are therefore significant, with an unprecedented demand for food and non-food agricultural products [FAO 16]. This is in addition to the expected effects of climate change (increasing temperatures, intensification of extreme events and increase in drought occurrence) on agricultural production and ecosystem biodiversity. In this context, mapping of the land cover/land use and its changes over time is an essential tool in order to (1) improve our understanding of spatial and temporal dynamics regarding cultivated and uncultivated landscapes, and (2) to manage and plan these landscapes.

Land cover mapping relies mainly on the classification of satellite images using supervised techniques (guided by the user to specify the land cover classes of interest). However, methods usually used in northern countries are ill-suited in an African context that is characterized by a high spatio-temporal variability of environmental conditions and agricultural practices: small fields, cloudiness during the crop season, quasi-simultaneous growing periods of cropped and natural vegetation, etc. [AFD 16]. The ESA's (European Space Agency) Sentinel-2 (S2) images available since 2015 and providing images at a high temporal (5-day revisit

Chapter written by Louise Leroux, Luca Congedo, Beatriz Bellón, Raffaele Gaetano and Agnès Bégué.

for S2A and S2B) and spatial (10 m) resolutions, offer new opportunities for land cover monitoring using remote sensing in Africa. In this chapter, a methodology for land cover mapping based on S2 and the Semi-Automatic Classification[1] plugin[2] (SCP) for QGIS [CON 16] is presented. We take as a case study the region of Yilou, located in the north of Ouagadougou in Burkina Faso.

4.2. Workflow for land cover mapping

Land cover mapping is based on the analysis of two S2 images acquired at the beginning (May) and at the end (November) of the cropping season in order to improve the spectral discrimination between classes of natural and cultivated vegetation. Figure 4.1 presents the general workflow that is divided into four main steps:

1) SCP installation and image download.

2) Pre-processing (radiometric correction, clipping of the study area, creation of spectral indices).

3) Supervised classification (creation of the training sites, assessment of the spectral signatures and classification).

4) Assessment of the classification accuracy and post-processing.

4.2.1. *Introduction to SCP and S2 images*

4.2.1.1. *Semi-Automatic Classification plugin*

SCP is a free and open-source plugin developed by Luca Congedo allowing the implementation of Semi-Automatic Classification based on a variety of satellite images such as MODIS, Landsat or S2. In addition to the classification module, it integrates many tools for pre-processing images (downloads, radiometric corrections, etc.) and post-processing (merging of classes, accuracy assessment, conversion from a classification raster to vector shapefile, etc.) making the SCP a comprehensive tool throughout the workflow for land cover mapping [CON 16]. A user manual and tutorials dedicated to the plugin are available on the website "From GIS to Remote Sensing"[3]. The plugin is available from the official repository of

---

1 Free and open-source plugin: https://fromgistors.blogspot.com/p/semi-automatic-classification-plugin.html.
2 Plugin in computing, also called extension, is a software component that adds a specific feature to an existing computer program (according to Wikipedia).
3 https://fromgistors.blogspot.com.

QGIS plugins and requires the default installation of GDAL, OGR, Numpy, SciPy and Matplotlib during QGIS standard installation.

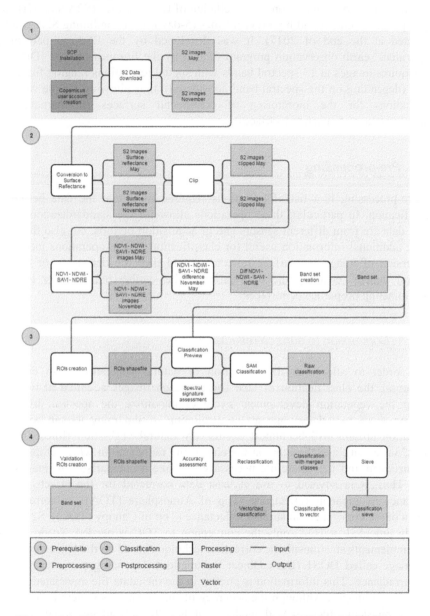

Figure 4.1. *Workflow for land cover mapping using SCP for QGIS. For a color version of the figure, see www.iste.co.uk/baghdadi/qgis2.zip*

4.2.1.2. S2 images

The S2 mission is based on a constellation of two satellites (S2A and S2B) with high spatial resolution and high revisit times (5-day revisit combining S2A and S2B expected at the end of 2017). It was developed by the ESA as part of the Copernicus[4] earth observation program funded by the European Union [DRU 12]. S2 acquires images in 13 spectral bands with spatial resolutions ranging from 10 to 60 m (depending on the spectral band considered), thus allowing a wide variety of applications for the monitoring of continental surfaces (in particular the characterization of vegetated surfaces).

4.2.2. Pre-processing

Pre-processing is a list of operations required to format the data before any classification. In particular, these operations allow for the standardization of data when data are from different sensors and/or acquisition periods, but also the ability create additional information useful for classification. These operations include the conversion of images to surface reflectance and then the derivation of different spectral indices, enabling the information to be synthesized in order to reveal discriminating properties of surfaces.

4.2.2.1. Conversion to surface reflectance

In order to allow the discrimination of natural vegetation from cultivated vegetation, the classification of land use relies on images acquired at two dates during the vegetation development cycle to maximize the spectral differences between these two dates: one at the beginning of the rainy season where the vegetation present will be almost exclusively natural vegetation (forest, riparian forest, savannah), and the other at the end of the rainy season where crops are not yet harvested and it is therefore necessary to make a comparison of images over time. Thus, it is advised to use at least data corrected for the effects of solar irradiance in apparent reflectance (Top of Atmosphere (TOA)) or corrected for effects of the atmosphere in surface reflectance (Top of Canopy (TOC)). S2 data are already provided in TOA, only the conversion to TOC is needed. To do this, the SCP implements an atmospheric correction technique based on the objects present in the image called DOS1 (Dark Object Subtraction) depending particularly on the solar irradiance. This information is present in the metadata file associated with the downloaded images. Although the accuracy of this method is generally lower than that of physical process-based methods, it has the advantage of not requiring

---

4 http://www.copernicus.eu.

measurements of atmospheric conditions. However, this method only corrects diffusion effects; it is therefore precise only in the visible domain but unreliable in the near-infrared and infrared domains dominated by the absorption of water vapor.

4.2.2.2. Clipping over the study area

S2 tiles cover a square area with sides of 110 km^2 (representing ~3 GO), thus the full classification process can be extremely time-consuming, especially as we include additional information (creation of spectral indices as the Normalized Difference Vegetation Index [NDVI]) in the initial dataset in order to improve classification (see section 4.2.2.3). Therefore, it makes sense to restrict the analysis to the extent of the study area.

4.2.2.3. Spectral indices

In order to highlight certain surface properties that are not visible on the original channels, new channels, or neo-channels, are created from the linear combination of reflectances in different wavelengths. Thus, there are many vegetation indices taking the form of very basic expressions (two-band ratio) or more complex expressions based on differences in spectral behavior of vegetation and soil, particularly marked in the red and near-infrared bands.

Among these indices, the most widely used and known is the NDVI [ROU 74, TUC 79], which is based on the reflectance in the red and the near-infrared whose difference increases with the density of green leaves and therefore with the chlorophyll concentration of the canopy. It is therefore a good indicator of the amount of green vegetation.

$$NDVI = \frac{\rho NIR - \rho R}{\rho NIR + \rho R} \hspace{4cm} [4.1]$$

where ρNIR is the reflectance in the near-infrared and ρR the reflectance in the red. The NDVI is a normalized index with values ranging from −1 to 1. For bare soils, NDVI values are close to 0 (due to small differences between the red and the infrared); for vegetated areas, NDVI values can vary from 0.1 to 0.8–0.9 (very dense and green canopy). However, NDVI is sensitive to the type of soil (for sparse vegetation cover), to atmospherics effects and viewing conditions and tends to saturate for dense vegetation.

To take into account the effects of soils on NDVI values, particularly for areas characterized by dense vegetation cover, the SAVI (Soil-Adjusted Vegetation Index [HUE 88]) has been developed by modifying the NDVI. The SAVI is structured similarly to the NDVI but with the addition of a correction factor, noted L, considering the soil–vegetation interactions:

$$SAVI = \frac{\rho NIR - \rho R}{(\rho NIR + \rho R + L)} \times (1 + L)$$ [4.2]

where ρNIR is the reflectance in the near-infrared (NIR), ρR the reflectance in the red and L the soil brightness correction factor. The value L varies with the amount of green vegetation, with values close to 0 in densely vegetated areas and values close to 1 for areas with no green vegetation. A value of 0.5 for L is generally used since it mitigates the influence of soil brightness. As for NDVI, values are ranged from -1 to 1.

Another simple index is the Normalized Difference Water Index (NDWI [GAO 96]), which is a construction similar to that of the NDVI but using a short wave infrared wavelength (water absorption peak) instead of a red band. NDWI values are comprised between -1 and 1 according to changes in the water content of leaves. NDWI is used to monitor vegetation in drought affected areas:

$$NDWI = \frac{\rho NIR - \rho SWIR}{\rho NIR + \rho SWIR}$$ [4.3]

where ρNIR is the near-infrared reflectance and $\rho SWIR$ is the reflectance in the short wave infrared.

Finally, for multispectral data such as S2 images, indices similar to NDVI but based on red-edge bands can be derived. Red-edge is the transition zone in the vegetation reflectance spectrum between the red and the NIR and therefore allows a better consideration of differences in chlorophyll absorption between vegetation cover and development conditions. Thus, the Normalized Difference Red-Edge Index (NDRI) is similar to NDVI but uses red-edge reflectance instead of the red reflectance:

$$NDRE = \frac{\rho NIR - \rho RE}{\rho NIR + \rho RE}$$ [4.4]

where ρNIR is the reflectance in the near-infrared and ρRE is the reflectance in the red-edge.

In order to take into account the dynamics of vegetation, channels of changes between the two dates can also be created by computing difference in the values of the spectral indices of November and May. By construction, values range from -2 to 2. For example, pixels with NDVI difference values close to 0 will indicate invariant land cover classes (e.g. bare soil or water or forest), while pixels with values tending to extremes will indicate changes between the two dates.

4.2.2.4. *Creation of the band set*

Once the spectral bands are converted into TOC, and the new spectral indices are created, the last pre-processing step consists of creating a group of layers or a multiband image containing all the bands necessary for the classification process, also called a "band set". It is therefore necessary to create a single dataset from the set of initial bands and spectral indices. This step also makes it possible to display combinations of bands (or color composite images) to facilitate visual discrimination of the different elements in the image (Figure 4.2).

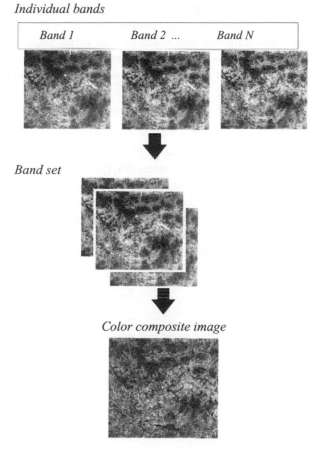

Figure 4.2. *Creation of a band set. For a color version of the figure, see www.iste.co.uk/baghdadi/qgis2.zip*

4.2.3. *Land cover classification*

The classification of a satellite image, in a broad sense, is defined as the classification of all pixels of the image within homogeneous groups (or classes) of land cover (nomenclature). Within each of these classes, the pixels have common spectral characteristics.

There are two main types of classification: the unsupervised classification where pixels are grouped according to their own structure and without prior intervention of the user on the definition of the classes, and the supervised classification where we seek to group pixels according to their spectral resemblance to reference objects representative of the land cover classes and defined *a priori* by the user. Supervised method land cover classification is thus based on four major steps: (1) creation of reference objects or training sites, (2) definition and analysis of spectral signatures of training sites, (3) classification and (4) accuracy assessment of the classification.

4.2.3.1. *Creation of ROIs*

After having determined the type and number of classes to be represented, the creation of training sites (or Region of Interest [ROI]) allows us to build a training database on which the classification algorithm is going to be trained to determine rules of discrimination between the different land cover classes. This step is based on the drawing of polygons on the representative image of each land cover class, chosen by taking each sample representative of intra-class variability (e.g. for a "Water Bodies" class, it would be necessary to take samples from rivers, lakes, etc.) (Figure 4.3).

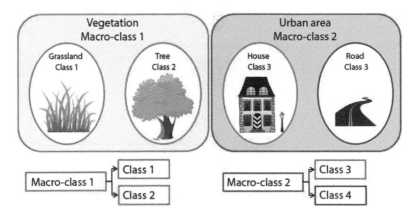

Figure 4.3. *Illustration of intra-class variability (adapted from [CON 16]). For a color version of the figure, see www.iste.co.uk/baghdadi/qgis2.zip*

In the training database:

– we must have training sites of sufficient size (area) to consider class-specific properties;

– each class must be represented by several training sites distributed evenly over the image;

– it is advisable to have a minimum of 10 training sites per class to ensure that the diversity of spectral properties of each class is represented in the training database;

– each training site must be as homogeneous as possible.

However, it should be noted that a training database obtained by photo-interpretation has a certain number of limitations, in particular that of being subjective and dependent on operator expertise. Therefore, the quality of the training database may in turn have an impact on the quality of the classification. This is why, as far as possible, it is better to have a training database obtained from a field campaign.

4.2.3.2. *Classification preview and assessment of spectral signatures*

The qualitative assessment of the training database quality is provided by the classification preview. A classification is created based on the training sites. It is possible to identify pixels that have been misclassified or forgotten in a class, meaning that the sampling is not representative of the spectral variability of this class. This step is therefore not the final classification but a temporary classification carried out on a portion of the image, which makes it possible to adjust the training sites.

In addition to the classification preview, the quality of the training database is assessed by analyzing the spectral separability (or spectral distance) between the different classes. If classes are weakly separable, the risk of confusion between these classes in the final classification is great. The objective of this step is (1) to verify the homogeneity of the sites in the same class, and (2) to check inter-class separability (Figure 4.4).

There are different measures of spectral distances that are chosen according to the specific classification algorithm:

– the Jeffries–Matusita distance, usually used for a Maximum Likelihood classification;

– the Spectral Angle, for a Spectral Angle Mapper (SAM) classification;

– the Euclidean distance, adapted for Minimal Distance classification;

– the Bray–Curtis similarity, which allows analyzing the similarity between two given samples.

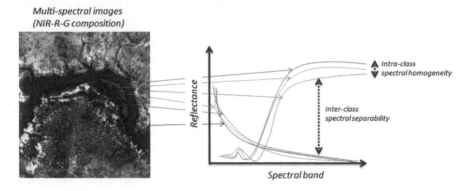

Figure 4.4. *Spectral signature and separability. For a color version of the figure, see www.iste.co.uk/baghdadi/qgis2.zip*

The training database creation step (section 4.2.3.1) and the pre-visualization/analysis of the classes' separability (section 4.2.3.2) are repeated as many times as necessary (iterative process). This is often the longest and tedious step in the classification process.

4.2.3.3. *Classification*

The classification step concerns the extrapolation of the entire image of the previously identified training sites, and for which a thematic land cover class could be attributed via a classification algorithm (model). The algorithm classifies each object (or pixels) present in the image by comparing its spectral characteristics with those of the reference objects in the training database. There are three main types of supervised classification: (1) metric, based on a unit of distance (e.g. Minimal Distance, SAM), (2) arithmetic (e.g. Hypercube) and (3) statistical (e.g. Maximum Likelihood, Random Forest).

For land cover mapping in regions characterized by high variability and heterogeneity of landscape features, as is the case in northern Burkina Faso, a SAM classification [KRU 93] is adapted because it allows us to group together pixels belonging to the same class but having different reflectances (typically at a date *t* plots can have different physiognomies according to the crop management practices). SAM determines the similarity between two spectral signatures by calculating the angle between a pixel and the spectral space within the training database and assigns it to the nearest class [KRU 93].

4.2.4. *Classification accuracy assessment and post-processing*

The final land cover mapping step consists of (1) the assessment of the classification accuracy and (2) post-processing to improve the classification.

4.2.4.1. *Classification accuracy*

Assessment of the classification accuracy is a fundamental step in the process because it allows quantifying the quality of the map obtained and the possibility to carry out additional processing (e.g. the merging of classes, sampling of new training sites, etc.) to improve the latter and validate it. To do this, the classification is compared with reference data (independent of those used to perform the classification) by means of a confusion matrix.

Two sampling designs can be implemented: stratified or random [GIR 10]. Ideally, the testing dataset should come from field observations or, alternatively, from photo-interpretation of satellite images. The reference data (or testing sites) are obtained in the same way as the training sites except that here the sites will be used here to assess *a posteriori* the quality of the classification. This new sampling must therefore be representative of the whole image.

The confusion matrix allows us to compute different statistics. The overall quality indices of the classification are as follows:

– The overall accuracy, which corresponds to the diagonal of the matrix and gives a well classified proportion of objects.

– The kappa index, which varies between 0 and 1 and expresses the reduction of the error obtained by the classification compared to a random classification [CON 91].

For each class, quality indices are as follows:

– The user's accuracy, which corresponds to the proportion of objects (or pixels) correctly, classified by comparing it to the reference (in column). The associated error (one-user accuracy) is the error of commission and gives an indication of the overestimation of the number of objects belonging to the different classes.

– The producer's accuracy is the proportion of reference objects correctly classified (in line). The associated error (one-producer accuracy) is the error of omission and gives an indication of the underestimation of the number of objects belonging to the different classes.

– The kappa index is an estimator of quality that takes into account both line and column errors. It allows both an overall assessment but also one at the level of each class.

4.2.4.2. *Post-processing*

4.2.4.2.1. Merging classes

Analyzing the confusion matrix can identify classes with strong confusion (strong spectral resemblance) and for which grouping may be necessary. In this case, similar classes are merged.

4.2.4.2.2. Filtering

In the final classification, isolated pixels (or small groups of pixels) can generally be observed. These isolated pixels can have an impact on the accuracy of the classification, and it may therefore be useful to delete them. A filter is then applied to replace the pixels isolated by the majority class observed in the pixels included in a defined neighborhood window (Figure 4.5).

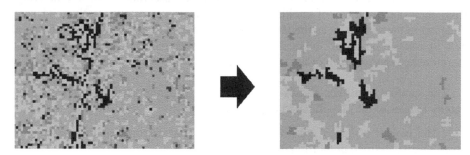

Figure 4.5. *Impact of a filter using a 4×4 pixel window. For a color version of the figure, see www.iste.co.uk/baghdadi/qgis2.zip*

4.2.4.2.3. Conversion to shapefile

Finally, the last step, but not always necessary, is vectorization. This step consists of converting raster data into vector data in the shapefile format. The attribute table of the vector file then has an attribute relating to the land cover classes.

4.3. Implementation with QGIS and the plugin SCP

After having the overall workflow for the land cover classification from satellite images (here S2 images), an example of application in QGIS was proposed, taking as a case study the region of Yilou, located in the province of Bam, in Burkina Faso for the year 2016. To do this, the SCP was used for the entire process [CON 16].

4.3.1. *Software and data*

4.3.1.1. *Software prerequisites*

Land cover mapping is based on basic QGIS functionalities (version 2.18) and the SCP. The installation of the plugin is done via the menu dedicated to plugin management.

Process	Practical implementation in QGIS
1. SCP plugin installation	In the **main menu toolbar**: • Click **Plugins > Manage and Install plugin**. In the **All** menu: • Browse and select **Semi-Automatic Classification Pl > Install plugin**. Verify the installation: • Click **Plugins > Manage and Install plugin > Installed**; • Or in the main menu toolbar: **SCP**.
2. SCP settings	Configuration of the available RAM. It is advised to use half of the available RAM. In the **main menu toolbar**: • Click **SCP > Settings > Processing**; • Set **Available RAM (MB)** to the targeted value, for instance 4096 for 4 GO of available RAM.

Table 4.1. *Software prerequisites*

4.3.1.2. *Creating a Copernicus account*

To download S2 images through the SCP plugin, the creation of a Copernicus account is required. The account is created via the Sentinel Scientific Data Hub[5] by clicking on the SIGN UP option at the top right hand corner of the window.

---

5 https://scihub.copernicus.eu/dhus.

4.3.1.3. *Data download*

The data used in this chapter are TOA S2 images, corrected geometrically with a Digital Elevation Model (level 1C)[6]. As part of this chapter, we use SCP to download the S2 images. However, these images can also be downloaded directly from the ESA-Copernicus Web portal.

In the semi-arid regions of sub-Saharan Africa, the growing season of both natural and cultivated vegetation is almost synchronized with the rainy season. It is therefore difficult to differentiate natural vegetation from crops using only one date. This is why this exercise uses two images acquired at the beginning of the rainy season (May) and at the end of the rainy season (November) to improve the identification of natural vegetation, which tends to start growing as soon as the first significant rains occur, instead of crops that grow later (sowing with the onset of rain).

Process	Practical implementation in QGIS
1. Data download	In the **main menu toolbar**: • Click **SCP > Download images > Sentinel-2 download**. In the **Login Sentinels** menu: • Give the login and the password of your Copernicus user account (see section 4.3.1.2). In the **Search Area** menu: • Give the geographical coordinates of the study area **UL X (Lon): -1.71, UL Y (Lat): 13.32, LR X (Lon): -1.71, LR Y (Lat): 12.97**. In the **Search** menu: • Set the time period between May 1, 2016, and November 30, 2016; • Set up **Max cloud coverage (%)** to **10%**; • Click **Find** and images with corresponding criterions will then be displayed in the **Sentinel images** menu. Pre-visualizing corresponding images: • Select images S2A_OPER_MSI_L1C_TL_SGS__20160515T175104_A0 04681_T30PXV and S2A_OPER_MSI_L1C_TL_SGS__20161111T155847_A0 07255_T30PXV. • Click the icon .

6 https://sentinel.esa.int/web/sentinel/user-guides/sentinel-2-msi/product-types/level-1c.

	In the **Download** menu: • Uncheck **Preprocess images** and **Load band in QGIS**; • Launch download by clicking the icon 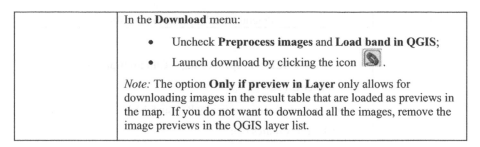. *Note:* The option **Only if preview in Layer** only allows for downloading images in the result table that are loaded as previews in the map. If you do not want to download all the images, remove the image previews in the QGIS layer list.

Table 4.2. *Downloading data with SCP*

4.3.2. *Step 1: data pre-processing*

4.3.2.1. *Conversion to surface reflectance*

Process	Practical implementation in QGIS
1. Conversion to surface reflectance	In the **main menu toolbar**: • Click **SCP > Preprocessing > Sentinel-2**. In the **Sentinel-2 conversion** tab: • Select the directory containing the Sentinel-2 bands; • Check **Apply DOS1 atmospheric correction** and uncheck **Create Band set and use Band set tools**. • Launch conversion by clicking the icon 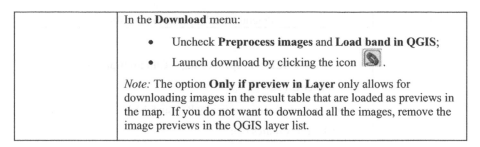. *Note:* The required information for the conversion are contained in the metadata file *MTD_SAFL1C*.xml, automatically read by the plugin. *Note:* The process has to be done for each date separately.

Table 4.3. *Conversion to surface reflectance*

4.3.2.2. *Clipping the data over the study area*

A clipping of images over the extent of the study area is applied. A step of reducing the size of the images is applied in order to obtain images that correspond to the area of interest. Furthermore, of the 13 bands in the S2 data (see section 4.2.1.2), only bands B3, B4, B6, B7, B8 and B11 will be used in the remainder of the study.

Process	Practical implementation in QGIS
1. Clipping images	In the **main menu toolbar**: • Click **SCP > Preprocessing > Clip multiple rasters**. In the **Raster list** tab: • Select images starting by RT_ and the 10 Sentinel-2 bands. In the **Clip coordinates** tab: • Set the geographical coordinates of the study area **UL X: 635529, UL Y: 1475323, LR X: 683489, LR Y: 1432381;** • Define a prefix for the output images in **Output name prefix** (ex. Resize); • Launch clipping by clicking the icon ; • In the QGIS layer list, clipped raster are displayed, the others can be removed. *Note:* It is also possible to clip data directly from a shapefile by checking the option **Use shapefile for clipping**. *Note:* The process has to be done for each date separately.

Table 4.4. *Reducing the size of the images to have images that correspond to the area of interest*

4.3.2.3. *Spectral indices calculation*

For each of the two dates, four neo-channels are computed, these are: NDVI, NDWI, SAVI and NDRI. Then, once the neo-channels are created for each of the two dates, an image containing changes in spectral indices values between the two dates is created (e.g. NDVI November to NDVI May), allowing for a better discrimination between classes. The computation of the spectral indices is automated through a batch.

Name	Formula	Formula for Sentinel-2
NDVI	$\dfrac{NIR - Red}{NIR + Red}$	$\dfrac{B08 - B04}{B08 + B04}$
NDWI	$\dfrac{NIR - SWIR}{NIR + SWIR}$	$\dfrac{B08 - B11}{B08 + B11}$
SAVI	$\dfrac{NIR - Red}{(NIR + Red + 0.5)} \times (1 + 0.5)$	$\dfrac{B08 - B04}{(B08 + B04 + 0.5)} \times (1 + 0.5)$
NDRI	$\dfrac{RE(780) - RE(730)}{RE(780) + RE(730)}$	$\dfrac{B07 - B06}{B07 + B06}$

Table 4.5. *Formulas of the used spectral indices*

Process	Practical implementation in QGIS
1. Data loading	**In QGIS**, load all the needed layers (previously clipped) for each date: **B03, B04, B06, B07, B08, B11**.
2. Opening the batch console	In the **main menu toolbar**: • Click **SCP > Batch**. In the **Function** option: • Select the function **!working_dir!** to set the working directory. The text !working_dir!;" is added to the batch window. Complete the path to the directory containing images: !working_dir!;'D:\Tempo\QGIS_Chap_LC\Projet\PreProcess'
3. Spectral indices calculation, NDVI example	In the **Function** option: • Select the **band_calc** function to define the function in order to compute NDVI. The text band_calc;expression : ";output_raster_path : ";extent_same_as_raster_name : ";align : 1;extent_intersection : 1;set_nodata : 0;nodata_value : 0 is added to the batch window. • In the **band_calc** function, replace: expression:" by expression:'(("Resize_RT_S2A_OPER_MSI_L1C_TL_S GS__20160515T175104_A004681_T30PXV_B08"-"Resize_RT_S2A_OPER_MSI_L1C_TL_SGS__20160515T 175104_A004681_T30PXV_B04")/("Resize_RT_S2A_OPE R_MSI_L1C_TL_SGS__20160515T175104_A004681_T30P XV_B08"+"Resize_RT_S2A_OPER_MSI_L1C_TL_SGS__ 20160515T175104_A004681_T30PXV_B04"))'

	• In the **band_calc** function, replace output_raster_path : " by output_raster_path : '!working_dir!\S2A_OPER_MSI_L1C_TL_SGS__20160515 T175104_A004681_T30PXV\Resize\Resize_RT_S2A_OPE R_MSI_L1C_TL_SGS__20160515T175104_A004681_T30P XV_NDVI.tif'
	• In the **band_calc** function, remove:extent_same_as_raster_name : ";align : 1 and keep the default value in extent_intersection : 1;set_nodata : 0;nodata_value : 0
	• Repeat step 3 for the second date, changing the path to images (Figure 4.6).
4. Calculation of the image difference for NDVI	In the **Function** option:
	• Select the **band_calc** function to define a function for computing the NDVI difference. The text: band_calc;expression : ";output_raster_path : ";extent_same_as_raster_name : ";align : 1;extent_intersection : 1;set_nodata : 0;nodata_value : 0
	is added to the window batch;
	• In the **band_calc** function, replace expression: " by expression: '("Resize_RT_S2A_OPER_MSI_L1C_TL_SGS__20161111 T155847_A007255_T30PXV_NDVI.tif"- "Resize_RT_S2A_OPER_MSI_L1C_TL_SGS__20160515T 175104_A004681_T30PXV_NDVI.tif")';
	• In the **band_calc** function, replace output_raster_path by : " par output_raster_path : '!working_dir!\S2A_OPER_MSI_L1C_TL_SGS__20161111 T155847_A007255_T30PXV\Resize\Resize_RT_S2A_OPE R_MSI_L1C_TL_SGS__20161111T155847_A007255_T30P XV_NDVIdif.tif';
	• In the **band_calc** function, remove extent_same_as_raster_name : ";align : 1 and keep the default value in extent_intersection : 1;set_nodata : 0;nodata_value : 0.
5. Calculation of NDWI, SAVI and NDRI and the associated difference image (Figure 4.7)	• Repeat steps 3 and 4 for each spectral index by modifying formula accordingly;
	• Launch calculation by clicking the icon ;
	• In the QGIS layer list, newly created images are displayed (Figure 4.6).

Table 4.6. *Calculation of the neo-channels with the batch option*

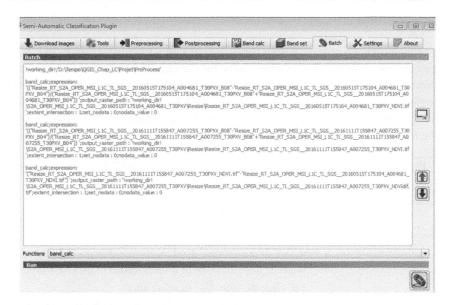

Figure 4.6. *Illustration of the batch interface for the calculation of spectral indices. Above you will find the NDVI calculation in May and November and the difference between the two months. For a color version of the figure, see www.iste.co.uk/baghdadi/qgis2.zip*

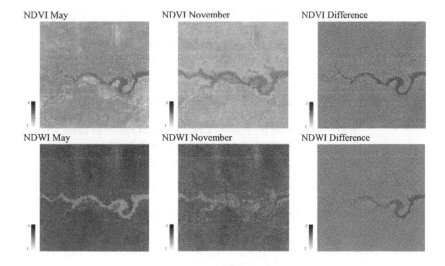

Figure 4.7. *NDVI and NDWI of May and November 2016 and the difference between the two dates for each index (November to May)*

4.3.2.4. *Creation of a band set*

The classification process within the SCP relies on a multiband image, that is, an image containing all the channels necessary for the classification process. An image containing the bands B03, B04, B08, B11, respectively, for the months of May and November and the spectral indices May NDVI, November NDVI, NDRIdif, NDVIdif, NDWIdif and SAVIdif will be created, see section 4.3.2.3). Other images can be removed from QGIS layer list.

Process	Practical implementation in QGIS
1. Creation of a band set	In the **main menu toolbar**: • Click **SCP > Band set**. In the **Band list** tab: • Select all images that should be included in the band set (14 in total). If there are only 14 images loaded in the QGIS layer list, click the icon 🔲 to select all. • Add selected images to the **Band set definition** window by clicking the icon ➕ . In the **Band set definition tab**: • Arrange bands as **May B03, B04, B08, B11, NDVI; November B03, B04, B08, B11, NDVI, NDRIdif, NDVIdif, NDWIdif, SAVIdif** using the ⬆ or ⬇ icons. They can also be ordered alphabetically with the 🔤 icon. • Leave the option **Quick wavelength settings** empty and select **band number** in the option **Wavelength unit**. • In the **Center wavelength** column verify that bands are numbered from **1 to 14**. If not, do it manually. This step is needed for the visualization and analysis of spectral signatures (see section 4.3.3.3). • Select the options **Create virtual raster of band set** and **Create raster of band set (stack bands)**. • Launch band set creation by clicking the icon 🖼 . *Note:* In cases where only original bands and one image are included in the band set, the Quick wavelength unit option is used to select the type of images to be entered and to automatically complete the Center wavelength column with the central values of the spectral bands.

Table 4.7. *Creation of a band set*

4.3.3. *Step 2: land cover classification*

Once all the data necessary for the mapping of land cover by supervised classification are formatted (preprocessing, calculation of the neo-channels and creation of the group of layers), the classification process with the SCP is then completed in four steps: (1) creation of training sites (ROI), (2) classification pre-visualizing, (3) analysis of the training site quality and lastly (4) classification.

4.3.3.1. *ROI creation*

As much as possible, the ROI should be representative of all land cover classes and intra-class heterogeneity. The land cover classes defined for our study area are:

– Agricultural Area;

– Bare Land;

– Forest;

– Urban Area;

   Water Bodies;

– Woodland;

Within the SCP, each land cover class is called Macro Class and is defined by a name (MC_Info) and by an ID (MC_ID). In our case study, the definitions are listed in Table 4.8.

MC_Info	MC_ID
Water Bodies	1
Forest	2
Agricultural Area	3
Urban Area	4
Woodland	5
Bare Land	6

Table 4.8. *Nomenclature used for classification*

For each of these Macro Classes, we will create several ROIs to have samples representative of the diversity within each land cover class (identified by a unique C_ID). The ROIs of the same Macro Class will therefore all have a unique MC_ID but different C_IDs, as in the example shown in Table 4.9.

MC_Info	MC_ID	C_ID	C_Info
Water Bodies	1	1	River
Water Bodies	1	2	River
Water Bodies	1	3	Pond
Water Bodies	1	4	Pond
Water Bodies	1	5	Lake

Table 4.9. *Illustration of several ROIs information for Macro Class "Water Bodies"*

Process	Practical implementation in QGIS
1. ROI file initialization	In **SCP Dock** (main interface of the SCP): • In the **SCP input** tab, click on the 🗁 icon for the **Training input** option to create a new training sites file; • Verify in **Input image** that the newly created file is selected. If necessary, click on the 🔃 icon to refresh the list. *Note:* The activation of the SCP Dock can be done with a right click in the toolbar and then in the section Panel, activate SCP Dock.
2. ROI creation	Image display: • During ROI creation, it is advised to regularly change the color composite. • In the SCP toolbar, in the RGB menu, specify the band numbers to display in the composition (e.g. 3-2-1 for NIR-Red-Green composition in May or 8-7-6 for the same composition in November). In the **SCP toolbar**: • Click on the icon to create a new ROI; • Draw a polygon and close it by right clicking. In the **SCP Dock**: • In the **ROI Creation** tab, set the **MC_ID** corresponding to the land cover class and the associated **MC_Info**. The **C_ID** increases automatically and a description can also be added. • Validate the ROI by clicking on the icon. The ROI and its spectral signature are added in the **ROI signature list** window from the **Classification dock** tab. • Repeat this step as many times as required until a representative sample is obtained (Figure 4.8).

Table 4.10. *ROI creation*

NIR-R-G May   NIR-R-G November

Water
bodies

Forest

Agricultural
areas

Urban
areas

Woodland
areas

Bare land

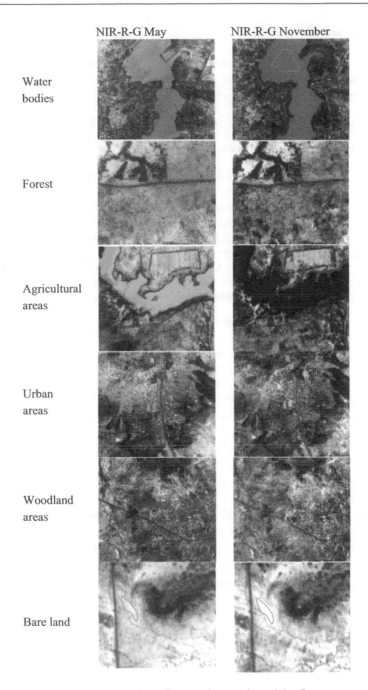

Figure 4.8. *Training sites. For a color version of the figure, see*
www.iste.co.uk/baghdadi/qgis2.zip

4.3.3.2. *Classification preview*

A first quality assessment of our training database and therefore of the final classification can be made from a preview of the classification. Here, the classification is made according to a SAM algorithm.

Process	Practical implementation in QGIS
1. Classification preview	In the **SCP Dock**: • In the **Macroclasses** tab, verify that all Macro Classes are there; • Eventually, the colors table can be modified by double clicking on each color or by clicking on the icon 🔲 **load qml** in the **Classification style** window in case a style has been previously created; • In the **Classification algorithm** tab, check **Use MC_ID**, and in the **Algorithm** window, choose **Spectral Angle Mapping**. In the **SCP toolbar:** • Click on 🔲 and click on an extract of the image where the classification algorithm will be applied; • The classification preview is displayed in the view windows and in the QGIS layers list.

Table 4.11. *Classification preview*

4.3.3.3. *Assessment of spectral signatures*

Spectral signatures are used by the classification algorithm to group pixels together according to their spectral similarity. The assessment of spectral signatures allows us to analyze the separability of the various Macro Classes. Several separability measures are implemented in the SCP. For the SAM classification, the Spectral Angle metric is the most appropriate:

$$0 < \theta < 90$$

$$Similar\ signature < \theta < Different\ signature$$

Process	Practical implementation in QGIS
1. Spectral signatures plot	In the **SCP Dock**: • In the **Classification Dock** tab, in the **ROI Signature list** window, select several spectral signatures for each Macro Class (Take ROI representative for each Macro Class, for instance for MC_Info=Water, take C_Info=River and Pond and Lake); • Click on the ⬚ icon for plot lines or on the ⬚ icon for a scatterplot; • **SCP: Spectral Signature Plot** window launch; • Each spectral signature is plotted as well as minimal and maximal values observed within ROI and this allows us to obtain information on intra ROI homogeneity (Figure 4.9).
2. Classes separability	In the **SCP** window: **Spectral Signature Plot**: • Select the spectral signature to be analyzed. The class separability is performed by comparing ROIs spectral signatures in twos. It is therefore advisable for one ROI per Macro Class to be taken. Click on the ⬚ icon, results obtained for the class separability measures appear in the **Spectral distances** window. Signatures of all ROIs are compared in twos and as a consequence there are as many separability measures as possible combinations of ROI.

Table 4.12. *Analysis of spectral signatures*

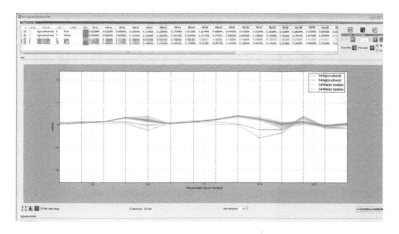

Figure 4.9. *Viewing of spectral signatures. Spectral signatures for two cultivated areas ROI and water bodies ROI are compared. For a color version of the figure, see www.iste.co.uk/baghdadi/qgis2.zip*

ROIs creation, classification preview and assessment of spectral signatures steps have to be repeated as much as needed until a good preview is obtained.

4.3.3.4. *Classification launch*

Process	Practical implementation in QGIS
1. Launch of classification	In the **SCP Dock**: • In the **Classification algorithm** tab, verify **Use MC_ID** is checked as well as **Spectral Angle Mapping** algorithm. • In the **Classification Output** tab, check **Classification report**. This will give an output file containing class statistics. Launch the classification by clicking the ⬛ icon. • The result of the classification is then displayed in the QGIS layers list (Figure 4.10).

Table 4.13. *Launch of classification*

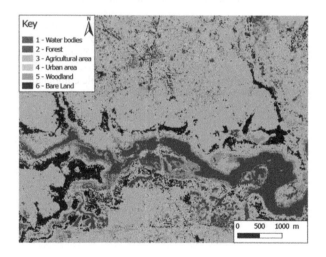

Figure 4.10. *SAM classification extract. For a color version of the figure, see www.iste.co.uk/baghdadi/qgis2.zip*

4.3.4. *Step 3: assessment of the classification accuracy and post-processing*

Five additional steps are needed to obtain the final classification: creation of testing sites and their photo-interpretation, assessment of the classification accuracy, post-processing (merging of similar classes and classification filtering), new accuracy assessment and conversion from a raster format to a shapefile format.

4.3.4.1. *ROIs creation for validation using a random sample and photo-interpretation*

The SCP automatically creates training/testing sites randomly on the extent of the image. These plots are then photo-interpreted in order to assign the corresponding MC_ID to them. Photo-interpretation can be done either from the S2 image, playing on the display of the different bands, or by using the OpenLayers plugin, which allows for the display of very high resolution images from Google or Bing.

Process	Practical implementation in QGIS
1. ROI file initialization	Be careful that the classification and the band set are displayed in the QGIS layers list. In the **SCP Dock**: • In the **SCP input** tab, click on the ▣ icon from the **Training input** option to initialize a new testing file. • Verify in **Input image** that the previously created band set is presented. Click on the ▣ icon to refresh the list. • In the **ROI creation** tab, check **Rapid ROI band** and specify the number 8 so that the testing sites are created automatically based on band 8, namely the NDVI for November.
2. Automatic and random creation of ROI	In **the SCP toolbar**: • Set the parameters for the automatic creation of ROI, with **dist** value at 0.010 (similarity between pixels to be merged, distance in radiometric unit), **Min** value (minimum surface of a unique ROI, expressed in number of pixels) at 20 (here 20 pixels) and **Max** value (the maximal width of a square in which the ROI is comprised and expressed in number of pixels) at 100 (here 100 pixels). *Note:* These parameters are only valid when the Rapid ROI band option is selected and the dist value has to be changed according to the band selected for the Rapid ROI creation. The default is the selection of the first band. In the **main toolbar**: • Click **SCP > Tools > Multiple ROI creation**. In the **Create random points (ROI)** menu: • Give the number of points that should be created with the **Number of points** option (for instance 300).

	• Check **min distance** to set a minimal geographical distance between each ROI (expressed according to the spatial unit of the image). Set the value to 1000 (i.e. 1000 m). • Click **Create points** ▣ . The list of created ROIs appears in the **Point coordinates and ROI definition** menu. In the **Run** menu: • Uncheck **Calculate sig** (option that allows us to compute spectral signatures of ROI); • Launch the creation of validation ROI by clicking on the icon 🖾 . *Note:* Setting a minimum distance can result in fewer points than the number defined in Number of points.
3. ROI photo-interpretation	In the **SCP Dock**: • Double click on the first ROI in the ROI list within the **ROI signature list** window from the **Classification dock** tab; • Set the corresponding **MC_ID** in the column **MC_ID** using the same **MC_ID** as those used for the testing ROI (see section 4.3.3.1); • Repeat this step for each ROI in the list.

Table 4.14. *ROIs creation for validation using a random sample*

4.3.4.2. *Classification accuracy before post-processing*

The accuracy of the classification is based on the testing ROI previously created. It results in a confusion matrix and several statistics: overall accuracy and kappa index, user accuracy, producer accuracy and kappa index for each of the land cover classes.

Process	Practical implementation in QGIS
1. Accuracy calculation	In the **SCP Dock**: • In the **Classification algorithm** tab, verify that **Use MC_ID** is checked. In the **main toolbar**: • Click **SCP > Postprocessing > Accuracy**. In the menu **Input** menu: • Select the classification to be assessed (**Select the classification to assess**); • Select the file containing testing ROI (**Select the reference shapefile or raster**);

- Click on the 🔄 icon to refresh the list;
- Select the field for MC_ID (**Shapefile field**);
- Launch the accuracy calculation by clicking on the 🖫 icon.

Outputs of the accuracy calculation are displayed in the **Output** menu. The overall accuracy is 64.5%, the kappa index 0.49 (Tables 4.16 and 4.17).

Note: It is also possible to manually create a testing file in QGIS (.shp), on condition of using the same identifiers (**MC_ID**) for the land cover classes.

Table 4.15. *Classification accuracy calculation*

Classification	ROI validation						
	1	2	3	4	5	6	Total
1	18290	0	1377	66	120	10	19863
2	9	70	243	6	1295	0	1623
3	32	24	22536	3047	13767	696	40102
4	17	0	1610	696	731	75	3129
5	39	56	8352	716	18177	62	27402
6	14	0	776	125	251	1302	2468
Total	18401	150	34894	4656	34341	2145	94587

Table 4.16. *Confusion matrix for the SAM classification*

Overall accuracy [%] = 64.56		
Class 1.0 Producer accuracy [%] = 99.4	User accuracy [%] = 92.1	Kappa hat = 0.90
Class 2.0 Producer accuracy [%] = 46.6	User accuracy [%] = 4.3	Kappa hat = 0.04
Class 3.0 Producer accuracy [%] = 64.6	User accuracy [%] = 56.2	Kappa hat = 0.30
Class 4.0 Producer accuracy [%] = 15.0	User accuracy [%] = 22.2	Kappa hat = 0.18
Class 5.0 Producer accuracy [%] = 53.0	User accuracy [%] = 66.3	Kappa hat = 0.47
Class 6.0 Producer accuracy [%] = 60.7	User accuracy [%] = 52.7	Kappa hat = 0.51
Kappa hat classification = 0.49		

Table 4.17. *Accuracy indices derived from the confusion matrix*

4.3.4.3. *Post-processing*

The confusion matrix analysis allows us to identify classes for which discrimination is difficult because of (for example) a small spectral difference. This is the case for class 2 (Forest) for which there is confusion with class 5 (Woodland). These two classes will be grouped together, meaning that for each pixel classified with MC_ID 2 the value 5 (MC_ID 5 for Woodland) will be given. Moreover, at the end of the classification, numerous isolated pixels are observed. A filter is applied to eliminate the isolated pixels thus also facilitating the conversion of the classification into a shapefile format.

Process	Practical implementation in QGIS
1. Classes merging	In the **main toolbar**: • Click **SCP > Postprocessing > Reclassification**; • Select the classification to be reclassed in the **Select the classification** option. In the **Values** menu: • Uncheck **calculate C_ID to MC_ID values** (option that allows us to automatically convert the C_ID values to the corresponding unique MC_ID value). • Click on the option **Calculate unique values** 🔲. The list of all Macro Classes (Different MC_ID) is calculated. Here a list of six values is given as outputs corresponding to the six previously defined MC_ID. • In the table, replace the MC_ID **2.0** (column **Old value**) by **5.0** in the column **New value**. In the **Symbology menu**: • Check the option **Use code from Signature list** and select **MC_ID**; • Launch the merging by clicking on the 🐌 icon.
2. Filtering	In the **main toolbox**: • Click **SCP > Postprocessing > Classification sieve** (option that allows us to filter the classification by replacing an isolated pixel with the majority values observed in its neighborhood); • Select the classification to be filtered with the option **Select the classification** (the reclassified classification in this case); • Set the **Size threshold** option to **6** (expressed in number of pixels, all patches of pixels smaller than 6 pixels will be replaced); • Set the **Pixel connection** option to **4** (4: in a 3×3 window, diagonal pixels are not considered connected, 8: in a 3×3 window, diagonal pixels are considered connected); • Launch the filtering by clicking on the 🐌 icon (Figure 4.11).

Table 4.18. *Class merging and filtering of isolated pixels*

Before post-processing   *After post-processing*

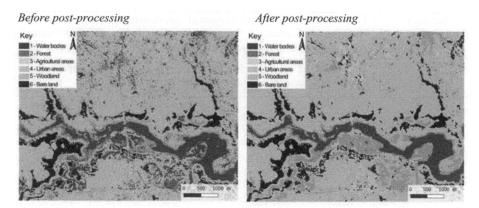

Figure 4.11. *Illustration of the effects of post-processing (class merging and filtering). For a color version of the figure, see www.iste.co.uk/baghdadi/qgis2.zip*

4.3.4.4. *Classification accuracy after post-processing*

When post-processing is done, the classification accuracy can be computed again to see the effect of these new processing steps on the final accuracy.

Process	Practical implementation in QGIS
1. Update of the ROI file	To compute the classification accuracy, the ROI file should be updated by replacing MC_ID for Forest (MC_ID = 2) by that of the Woodland class (MC_ID = 5) since these two classes have been merged previously. In the **QGIS layers list**: • Right click on the ROI testing file > **Open attribute table**; • Click on the icon to filter features using form ; • In the **Expression** tab, type "MC_ID"=2 and validate to select ROI corresponding to Forest Class; • Click on the icon to open the **Field calculator** ; • Check the option **Only update X selected features**; • Check the option **Update existing field** and select **MC_ID**; • In the **Expression** tab, write 5 in order to attribute the MC_ID of Woodland (MC_ID = 5) to all ROIs selected and previously classified as Forest; • Save modifications and quit tool editing mode .

2. Accuracy calculation	Follow the same steps presented previously in section 4.3.4.2, classification accuracy before post-processing but put the classification after post-processing in the option **Select the classification to assess**.
	Outputs of the accuracy calculation are displayed in the **Output** menu. The overall accuracy is now 68.5%, the kappa index 0.53.

Table 4.19. *Evaluation and accuracy of classification after post-processing*

4.3.4.5. *Conversion to vector*

Process	Practical implementation in QGIS
1. Conversion to vector	In **the main toolbar**: • Click **SCP > Postprocessing > Classification to vector**; • Select the classification to convert in vector in the option **Select the classification**. In the **Symbology tab**: • Check the option **Use code from Signature list** and select **MC_ID**; • Launch the conversion to vector by clicking the ⬛ icon. *Note:* Depending on the size of the area and the numbers of features, this may take a long time.

Table 4.20. *Vectorization*

4.4. Bibliography

[AFD 16] AFD-CIRAD, "Observation spatiale pour l'agriculture en Afrique : potentiels et défis ", available at: agritrop.cirad.fr/579494/1/12-notes-techniques.pdf, 2016.

[CON 16] CONGEDO L., "Semi Classification Plugin Documentation", available at: https://fromgistors.blogspot.com/p/semi-automatic-classification-plugin.html, 2016.

[CON 91] CONGALTON R.G., "A review of assessing the accuracy of classifications of remotely sensed data", *Remote Sensing of Environment*, vol. 37, no. 1, pp. 35–46, 1991.

[DRU 12] DRUSCH M., DEL BELLO U., CARLIER S. *et al.*, "Sentinel-2: ESA's optical high-resolution mission for GMES operational services", *Remote Sensing of Environment*, vol. 120, pp. 25–36, 2012.

[FAO 16] FAO, "La situation mondiale de l'alimentation et de l'agriculture 2016 : Change-ment climatique, agriculture et sécurité alimentaire", available at http://www.fao.org/3/a-i6030f. pdf, 2016.

[GAO 96] GAO B.C., "NDWI – A Normalized Difference Water Index for remote sensing of vegetation liquid water from space", *Remote Sensing of Environment*, vol. 58, pp. 257–266, 1996.

[GIR 10] GIRARD M.C., GIRARD C., *Traitement des données de télédétection: Environnement et ressources naturelles*, Dunod, Paris, 2010.

[HUE 88] HUETE A., "A soil-adjusted vegetation index (SAVI)", *Remote Sensing of Environment*, vol. 8, pp. 295–309, 1988.

[KRU 93] KRUSE F.A., LEFKOFF A.B., BOARDMAN J.W., *et al.*, "The spectral image processing system (SIPS) – interactive visualization and analysis of imaging spectrometer data", *Remote Sensing of Environment*, vol. 44, pp. 145–163, 1993.

[ROU 74] ROUSE J., HAAS R., SCHELL J., Monitoring the vernal advancement and retrogradation (greenwave effect) of natural vegetation, NASA/GSFC Type III Final Report, NASA/GSFC, Greenbelt, 1974.

[TUC 79] TUCKER C.J., "Red and photographic infrared linear combinations for monitoring vegetation", *Remote Sensing of Environment*, vol. 8, no. 2, pp. 127–150, 1979.

Detection and Mapping of Clear-Cuts with Optical Satellite Images

5.1. Definition and context

Clear-cutting refers to a logging practice in which all trees in a forest plot are cut down. This practice is regulated by French law, including the Forest Code. It requires applications for authorization (L124-5, L312-9, L312-10, R312-19 to 21) and measures to restore and rehabilitate forest stands (L124-6). Sanctions (L163-2, L362-1 and D312-22) can also be applied in the case of illicit or abusive logging, or if there is not any rehabilitation following a clear-cut.

The French Ministry of Agriculture, in charge of forest policies, has commissioned the development of a clear-cuts detection method with optical satellite images. This method, intended for various state services, is an effective method of pre-diagnosis. It gives forestry officers the means to target and optimize the field-based controls, including compliance with clear-cutting permissions and reforestation obligations.

The method has been adapted to the free and open-source software QGIS. It can be used by any GIS engineer or technician, without specific skills in remote sensing.

Chapter written by Kenji OSE.

The technical specifications expressed by the ministry are as follows:

Thematic content	Detection of clear-cuts between two consecutive years
Geographic accuracy	Minimum detection area of 1 ha with a planimetric accuracy of 5–10 m
Thematic accuracy	Omission (undetected clear-cut) and commission (false positive) errors less than 10% (based on area)
Projection system	Lambert 93 – French Geodetic Network 1993

Table 5.1. *Technical specifications for the mapping of clear-cuts*

5.2. Clear-cuts detection method

Mapping the clear-cuts relies on a change detection method based on two satellite images acquired between two consecutive years, during summer when trees show their leaves. Figure 5.1 describes the processing sequence. For ease of understanding, data processing is divided into seven main steps, from change detection to qualification of results:

1) image acquisition, calculation of the difference in vegetation indices between two consecutive years;

2) creation of raster masks for pixels outside the forest;

3) integration of raster masks;

4) qualification of forest changes and classification of detected clear-cuts;

5) export to vector format;

6) attribute data enrichment (degree of certainty, area, etc.);

7) statistical evaluation.

5.2.1. *Step 1: change detection – geometric and radiometric pre-processing*

The proposed approach consists of creating a land cover change image. First, optical satellite images acquired at different dates are converted, if necessary, into reflectance and then into vegetation indices. The subtraction of indices gives a mono-band image in which two main change types can be visualized: the transition from a vegetated state to a bare soil state or other, and vice versa.

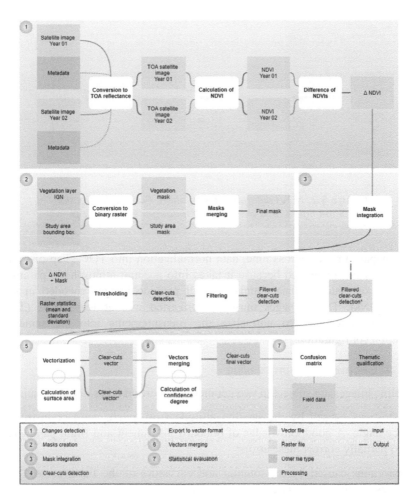

Figure 5.1. *Processing chain for clear-cuts mapping. For a color version of the figure, see www.iste.co.uk/baghdadi/qgis2.zip*

5.2.1.1. *Geometric and radiometric pre-processing*

The pre-processing of optical images aims to make images, possibly acquired by different satellites, comparable. Above all, images must be stackable and projected in a well-known spatial reference, for example Lambert 93-RGF93.

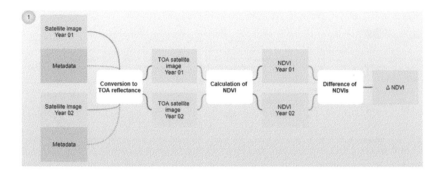

Figure 5.2. *NDVI difference calculation. For a color version of the figure, see www.iste.co.uk/baghdadi/qgis2.zip*

To compare images across time, data normalization into TOC (Top-of-Canopy) reflectance is recommended because it takes into account atmospheric parameters. However, most satellite data companies still provide images encoded in digital numbers (DNs). Then, images must be converted into at least TOA (Top-of-Atmosphere) reflectance to normalize the measurements into the same solar illumination configuration.

5.2.1.1.1. Geometric pre-processing

Satellite images are subject to two geometric distortion types during acquisition:

– distortions caused by the satellite attitude (roll, pitch, yaw) or instrument's incidence angle. A good knowledge of onboard instrument and flight parameters is useful to partially correct these errors;

– distortions caused by the rotation and curvature of the Earth as well as the relief. These factors are more complex to model.

For most of the satellite coverages provisioned by public programs such as Copernicus[1], Theia/Geosud[2], USGS/NASA EarthExplorer[3] or some Internet giants such as Google[4] and Amazon[5], satellite images have already been ortho-rectified. If so, just make sure the images have the same projection system.

---

1 https://scihub.copernicus.eu.
2 https://www.theia-land.fr; http://ids.equipex-geosud.fr.
3 https://earthexplorer.usgs.gov.
4 https://earthengine.google.com/datasets.
5 http://sentinel-pds.s3-website.eu-central-1.amazonaws.com.

Otherwise, image geometry has to be corrected (see [OSE 16]) using planimetric (e.g. an already ortho-rectified image) and altimetric (e.g. a digital elevation model [DEM]) reference data.

QGIS functionality to change image projection:
- System projection change: *Raster > Projections > Warp (Reproject)...*

5.2.1.1.2. Radiometric pre-processing

Radiometric pre-processing involves converting pixel values from DN – encoded into 8-bit (256 values) or 16-bit (65,536 values) – to reflectance (%). For the clear-cut detection method, only the TOA reflectance is described here.

This processing step takes into account several calibration parameters reliant on the satellite's sensor and the acquisition date (because solar illumination changes with seasons):

1) the sensor gain and offset (or bias);

2) the solar zenith angle;

3) the solar irradiance.

As most of these required parameters change for each spectral band of the same image [OSE 16], the calibration into TOA reflectance may be a tricky processing step. In addition, calibration formulas are often different according to sensors. In QGIS, the plugin "Geosud TOA Reflectance" automatically converts the images acquired by the satellites SPOT-5, SPOT-6/7, Pléiades, Rapideye and Landsat 8 to TOA reflectance. This tool avoids miscalculations and saves processing time.

QGIS functionality to convert pixel values from DN to TOA reflectance:
- TOA reflectance calibration: *Raster > Geosud TOA Reflectance > convert DN to reflectance*

Some satellite datasets no longer require radiometric pre-processing. New sensors such as Sentinel-2 or Landsat 8 supply TOA reflectance images by default. These images can be downloaded from public (Copernicus, USGS, etc.) or private (Google, Amazon, etc.) platforms. In addition, within the French Theia Land Data Centre (http://www.theia-land.fr), CNES (French Space Agency) and Cesbio (laboratory) have developed "Muscate", which is a processor that produces ready-to-use TOC reflectance data in near-real time with good cloud and cloud shadow masks. It is able to deliver corrected images from Landsat[6], SPOT[7], Formosat-2, Venµs and Sentinel-2[8].

As these images do not require geometric or radiometric corrections, it is possible to convert them to vegetation indices without any other pre-processing.

By definition, a clear-cut is a land cover change. It is therefore necessary to detect changes that have occurred between two dates, using vegetation indices in particular, and then to qualify these changes in order to retain only the forest plots that have passed from a wooded state to a bare soil state.

5.2.1.2. The Normalized Difference Vegetation Index (NDVI)

The NDVI makes it possible to observe and analyze vegetation cover on satellite images. This index reflects the very characteristic spectral signature of active vegetation. Due to leaf pigments, vegetation reflection is very low in the visible spectrum, particularly in the red band (0.6–0.7 µm), whereas it increases sharply in the near-infrared band (0.7–0.9 µm). Since bare soils have a small reflectance gap between red and near-infrared, the distinction between bare and covered soil is generally not a problem:

$$NDVI = \frac{\rho_{TOA}^{NIR} - \rho_{TOA}^{Red}}{\rho_{TOA}^{NIR} + \rho_{TOA}^{Red}} \tag{5.1}$$

where ρ_{TOA}^{NIR} is the reflectance of the near-infrared band and ρ_{TOA}^{Red} the reflectance of the red band.

The denominator is a normalization factor that partially compensates for the difference in surface reflectance related to the sun height or sensors' acquisition angles. By design, the NDVI varies between -1 and $+1$.

QGIS functionality to calculate NDVI:
- NDVI calculation: *Raster > Raster Calculator…*

5.2.1.3. NDVI difference and image of changes

The image of changes is the difference Δ_{NDVI} of the NDVIs calculated on two consecutive years:

$$\Delta_{NDVI} = NDVI_{D2} - NDVI_{D1} \tag{5.2}$$

where $NDVI_{D1}$ and $NDVI_{D2}$ are vegetation indices of dates $D1$ and $D2$.

---

6 http://spirit.cnes.fr/resto/Landsat.
7 https://spot-take5.org.
8 https://theia.cnes.fr/atdistrib/rocket/#/search?collection=SENTINEL2.

Δ_{NDVI} values vary between -2 and $+2$. Visually, the pixels that tend to white or black reflect land cover changes. Gray hues correspond rather to elements that are invariant over time.

Figure 5.3 illustrates the NDVIs difference:

1) At date $D1$, a forest has a high NDVI of 0.7. Inside is a bare soil plot (old clear-cut) with a low NDVI (0.1).

2) At date $D2$, the previous bare soil plot is plant-covered again (NDVI = 0.6). Two plots have undergone clear cuts, their NDVI are equal to 0.1 and 0.2.

3) With the difference of both NDVI layers ($NDVI_{D2} - NDVI_{D1}$), it is possible to identify three types of evolution:

i) the areas with little or no change have values close to 0;

ii) the areas that have passed from bare soil state to vegetation state have positive values, significantly different from 0, with a maximum of 2;

iii) the areas that have passed from vegetation state to bare soil state have negative values, significantly different from 0, with a minimum of -2.

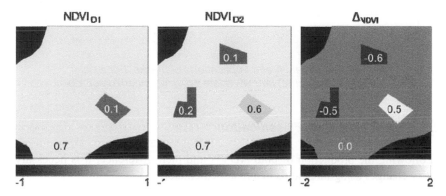

Figure 5.3. *NDVI on dates D1 and D2 and NDVIs difference*

QGIS functionalities:
- Rasters alignment: *Raster > Align Rasters...*
- NDVI and Δ_{NDVI} Calculation: *Raster > Raster Calculator ...*

5.2.2. *Steps 2 and 3: forest delimitation*

The second step is to hide pixels outside the study area and the forest areas. Two vector layers are converted into binary rasters and then merged. The first one allows the data to be cut along the study area boundaries (e.g. a "sylvo-ecoregion"). The second one, from the "vegetation" vector file from the Topo® IGN database or Forêt® IGN database (finer in terms of nomenclature), is dedicated to forest areas extraction. The resulting mask is therefore integrated into the NDVI difference produced in step 1 in order to only provide an image of changes on forest areas. Pixels outside the study area and forests are then coded as "no data". This step is mandatory because the detection of clear-cuts is based on the distribution statistics of pixels included in the wooded areas only.

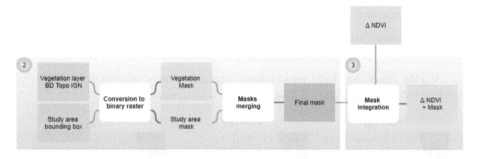

Figure 5.4. *Creation and applications of masks. For a color version of the figure, see www.iste.co.uk/baghdadi/qgis2.zip*

5.2.3. *Step 4: clear-cuts classification*

Following the production of both NDVIs and their difference, it is necessary to identify changes using thresholds in a raster calculator.

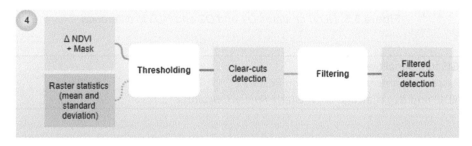

Figure 5.5. *Classification of changes. For a color version of the figure, see www.iste.co.uk/baghdadi/qgis2.zip*

In this method of thresholding and for the sake of genericity, the distribution of Δ_{NDVI} pixel values is assumed to follow a normal distribution. Moreover, a forest that has not changed may have slight NDVI variations over time, and its Δ_{NDVI} thus occupies a range of more or less one standard deviation around the mean. Values between minimum (-2) and the mean of Δ_{NDVI} values minus one standard deviation show the change from a wooded state at date $D1$ to a bare or slightly vegetated state at date $D2$:

$$request\ 1: [-2 < \Delta_{NDVI} < (m - \sigma)] \tag{5.3}$$

where m is the mean and σ the standard deviation of Δ_{NDVI} pixels values located in a forest.

This first request delivers a binary image in which positive pixels (coded in 1) correspond to areas assumed to be cut. This result is still imprecise and difficult to interpret because it assigns the same value to different types of cuts: recent or old cuts, certain or probable cuts, etc. It is therefore proposed to discretize values included in the interval of the first request, into three classes ordered according to a confidence degree. Thus, the more the value of a pixel moves away from the mean of Δ_{NDVI}, the stronger is the certainty that it is a recent clear-cut. Standard deviation measures the classes' amplitude. So, the expression is expressed as follows:

$$
\begin{aligned}
&class\ 1 = [(m - 2\sigma) \le \Delta_{NDVI} < (m - \sigma)] \\
request\ 2: \quad &class\ 2 = [(m - 3\sigma) \le \Delta_{NDVI} < (m - 2\sigma)] \\
&class\ 3 = [-2 \le \Delta_{NDVI} < (m - 3\sigma)]
\end{aligned}
\tag{5.4}
$$

From this second request results an image encoded in integers between 0 and 3. The value "no data" refers to any part of the image without forest, the value 0 to forest areas that have not been cut and the values from 1 to 3 correspond to the detections of clear-cuts with a low, medium or high degree of confidence. In practice, degree 1 most often contains false detections caused by atmospheric or topographic (shadows) effects.

QGIS functionalities:
- Thresholding: *Raster > Raster Calculator …*

5.2.3.1. *Filtering*

At the end of the classification, a raster is created. The values 1, 2 or 3 identify the clear-cuts. However, these data include groups of pixels whose surface area falls short of the specifications (minimum area of one hectare), or even isolated pixels. A filter is applied to eliminate these elements and reduce the number of entities when exporting in vector mode.

The GDAL Sieve Filter is applied to images. This tool eliminates raster objects whose surface area (in pixels) is less than the threshold specified by the operator, replacing their value with that of the largest neighbor object. Besides its size, a raster object is also defined according to its connectivity, which expresses the relationship of a pixel with its neighbors. A pixel with a connectivity of 4 is adjacent to any pixel sharing a same side; a pixel with a connectivity of 8 is adjacent to any pixel sharing a same side or a same vertex. The filtering thus affects the final shape of objects (Figure 5.6).

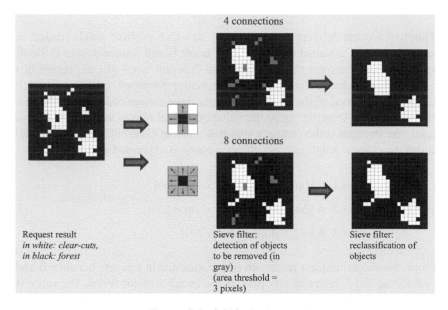

Figure 5.6. *QGIS sieving tool*

QGIS functionalities:
- Filtering: *Raster > Analysis > Sieve...*

5.2.4. *Steps 5 and 6: export in vector mode*

5.2.4.1. *Vector and attributes*

The images of the second request, optionally filtered (sieving), are converted into vector in shapefile format (ESRI). For each entity, the confidence degree is recovered as an attribute. The output attribute table is enriched with new fields, with at least the area in hectares. It is recommended to add other information such as false detections (caused, for example, by the presence of clouds).

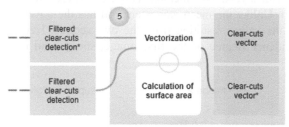

In case several images are needed to cover the study area on both dates.

Figure 5.7. *Vectorization. For a color version of the figure, see www.iste.co.uk/baghdadi/qgis2.zip*

Finally, entities less than 1 ha (area field) and errors found in the image are deleted or, failing that, excluded from cartographic renderings and statistical calculations.

QGIS functionalities:
- Vector conversion: *Raster > Conversion > Polygonize (Raster to Vector)...*
- Surface area calculation: *Open field calculator*

5.2.4.2. *Vectors superposition (optional)*

Should several images be needed to cover the study area on both dates, output vector layers can be gathered in a single file using geometric union (geoprocessing tool). It retains the attributes of input entities.

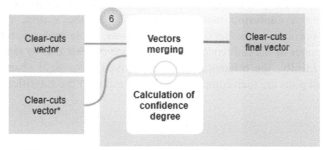

In case several images are needed to cover the study area on both dates.

Figure 5.8. *Vectors merging*

In the overlap areas, the intersected entities often have two different confidence degrees since satellite images are rarely acquired on the same date. It is therefore necessary to recalculate this indicator in order to synthesize the information coming from the input vector layers. Two methods (Figure 5.9) are available: the mean and the concatenation. Unlike the mean, concatenation has the advantage of ensuring a "traceability" of the values resulting from the two vectors. Here, the first number defines the confidence degree of the first image; the second number that of the second image.

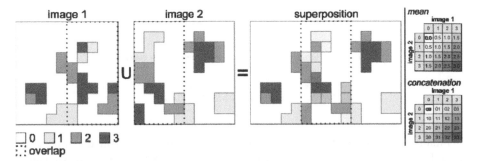

Figure 5.9. *Union of detected clear-cuts vectors. For a color version of the figure, see www.iste.co.uk/baghdadi/qgis2.zip*

QGIS functionalities:
- Processing Toolbox: *QGIS geoalgorithms > Vector overlay tools > Union*
- Surface area calculation: *Open field calculator*

5.2.5. *Step 7: statistical evaluation*

At the end of the processing chain, the quality of the clear-cuts mapping is evaluated using a confusion matrix (Table 5.2). In the matrix, each column represents the occurrence number of a classified class, while each row represents the occurrence number of the reference. The occurrence number can be replaced by the surface area. Ideally, the reference data are derived from field observation. The appraised forest plots are then compared with the results of the classification.

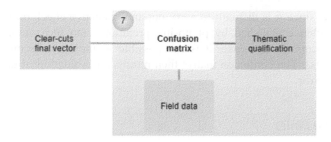

Figure 5.10. *Mapping quality of evaluation. For a color version of the figure, see www.iste.co.uk/baghdadi/qgis2.zip*

From this matrix, the following are calculated:

– overall accuracy refers to the proportion of well-classified entities (diagonal matrix);

– user accuracy, is the proportion of well-classified classification entities with respect to the reference. It reads by column;

– producer accuracy is the proportion of well-classified reference entities. It reads by row;

– omission and commission errors reflect the attribution of an incorrect label to an entity. More simply, they indicate whether the classification process tends to overestimate or underestimate the number (or surface area) of entities belonging to a land cover class.

		Classification				
		Clear-cuts	Other	Total rows	Producer accuracy (%)	Omission error (%)
	Clear-cuts	X_{11}	X_{12}	N_1	$100 \cdot \dfrac{X_{11}}{N_1}$	$100 \cdot \dfrac{X_{12}}{N_1}$
Reference	Other	X_{21}	X_{22}	N_2	$100 \cdot \dfrac{X_{22}}{N_2}$	$100 \cdot \dfrac{X_{21}}{N_2}$
	Total columns	M_1	M_2	N		
	User accuracy (%)	$100 \cdot \dfrac{X_{11}}{M_1}$	$100 \cdot \dfrac{X_{22}}{M_2}$			
	Commission error (%)	$100 \cdot \dfrac{X_{21}}{M_1}$	$100 \cdot \dfrac{X_{12}}{M_2}$			
	Overall accuracy	$100 \cdot \dfrac{X_{11} + X_{22}}{N}$				

Table 5.2. *Example of confusion matrix*

In practice, since it is difficult to carry out an exhaustive check on the field, the reference is often established by photo-interpretation from the vector file of the detected clear-cuts. In this case, it is not possible to measure the omission error (undetected cuts). The confusion matrix then returns an evaluation of user accuracy and omission error.

5.2.6. *Method limits*

Although the clear-cut detection method is applicable to a variety of datasets, the results may overestimate or omit some plots for at least two reasons:

– The vegetation index makes it easy to differentiate vegetation from bare soil surfaces. However, the state of bare soil following a clear-cut is brief. Vegetation takes over in a few months. Detection is therefore particularly sensitive to the satellite revisiting frequency over a certain place. The more spaced out acquisitions are, the greater the risk of omission is.

– The presence of clouds on the image systematically leads to false detections. Some providers deliver a cloud mask in addition to the image. The quality of this mask is also very variable.

To solve these problems, it is recommended to use images from systematic acquisition satellites such as Landsat 8 (revisits every 16 days) or the Sentinel-2 constellation (revisits every 5 days). These data are integrated into post-processing chains developed within the framework of different scientific programs (Copernicus, Theia, etc.). They are thus corrected for atmospheric disturbances (cirrus, aerosols, etc.) and benefit from cloud masks of fairly good quality.

5.3. Practical application

This section presents a practical application of the clear-cuts detection on a couple of Landsat 8 images acquired during summer 2015 and 2016, between Tours and Besançon (France).

5.3.1. *Software and data*

5.3.1.1. *Required software*

The clear-cut detection method uses basic functionalities of QGIS (version 2.16) to process raster and vector data. This choice allows anyone to apply processings without the need to install extensions. Of course, some steps of the method can be carried out with more efficient tools, in particular the Orfeo ToolBox (OTB) image processing library developed by CNES.

5.3.1.2. Input data

This exercise is based on open and free data, which can be downloaded from the Internet. These are Landsat 8 (Level 2A) reflectance images – processed by the Theia land services French Centre – and the 2012 CORINE Land Cover layer, accessible from the Observation and Statistics Service (SOeS) Website of the General Commissariat for Sustainable Development (CGDD).

5.3.1.2.1. Download of Landsat 8 images

This exercise uses two Landsat 8 images acquired on August 3, 2015, and August 14, 2016, on tile D0007H0006. Data can be downloaded from the following Weblinks:

– Landsat 8 – August 3, 2015

File name: LANDSAT8_OLITIRS_XS_20150803_N2A_France-Metropole D0007H0006.tar

Link: https://theia-landsat.cnes.fr/rocket/#/collections/Landsat/b9587432-31d7-542d-85ad-a2740bdb7d41

Figure 5.11. *Landsat 8 image acquired on August 3, 2015. For a color version of the figure, see www.iste.co.uk/baghdadi/qgis2.zip*

– Landsat 8 – August 14, 2016

File name: LANDSAT8_OLITIRS_XS_20160814_N2A_France-Metropole D0007H0006.tar

Link: https://theia-landsat.cnes.fr/rocket/#/collections/Landsat/7f0d72e3-9be4-51b6-ab71-4ca5602d8566

NB: To download the images, it is necessary to create an account in advance (free registration).

Figure 5.12. *Landsat 8 image acquired on August 14, 2016. For a color version of the figure, see www.iste.co.uk/baghdadi/qgis2.zip*

5.3.1.2.2. Download of CORINE Land Cover 2012

The CORINE Land Cover vector layer is available at the following address:

File name: CLC12_RBFC_RGF_SHP.zip

Link: http://www.statistiques.developpement-durable.gouv.fr/clc/fichiers/

NB: Download only the Great Region "Bourgogne, Franche-Comté".

5.3.2. *Step 1: creation of the changes image*

Compressed files distributed by Theia each have two images with different correction levels. For the exercise, only the images taking into account slope effects (ORTHO_SURF_CORR_PENTE) are used here.

To improve readability in the following sections, these images are renamed: LSAT_2015.tif and LSAT_2016.tif.

5.3.2.1. *Calculation of the NDVI*

NDVI is a normalized index with values ranging from –1 to 1. An NDVI close to 1 indicates active vegetation, close to 0 bare soils and below 0 wetlands or water bodies.

Step	QGIS handling

1. Calculation of NDVI	In **QGIS**:
	
	  • Open the image LSAT_2015.tif.
	
	In the **menu bar**:
	
	  • Click on **Raster > Raster Calculator…**
	
	In **Raster calculator**:
	
	  • Enter the following NDVI formula:
	
	("LSAT_2015@5" - "LSAT_2015@4") /
	
	("LSAT_2015@5" + "LSAT_2015@4")
	
	  • Name the output file: NDVI_LSAT_2015.tif.
	In the same way, calculate the NDVI of the next year (LSAT_2016.tif), and name the output file: NDVI_LSAT_2016.tif.

Table 5.3. *Calculation of NDVI*

5.3.2.2. *Creation of the change image*

The change image is simply the difference between the NDVIs on both dates. The result is a monochrome image. Visually, pixels that tend toward white or black reflect changes in land cover. The gray shades correspond rather to elements that are invariant over time.

Step	QGIS handling

1. Input data verification	In **QGIS**:
	
	  • Check for the following files:
	    o NDVI_LSAT_2015.tif
	    o NDVI_LSAT_2016.tif

2. Calculation of the change image	In the **menu bar**:
	
	  • Click on **Raster > Raster Calculator…**
	
	In **Raster calculator**:
	
	  • Enter the following expression:
	
	"NDVI_LSAT_2016@1" – "NDVI_LSAT_2015@1"
	
	  • Save the file as DIFF_NDVI_2015-2016.tif.

Table 5.4. *Calculation of the change image (difference of NDVIs)*

5.3.3. *Steps 2 and 3: creation, merging and integration of masks*

This step consists of removing any pixel that is located outside the forest zones or which has radiometric defects (saturation, cloudiness, etc.) from the analysis.

5.3.3.1. *Management of Theia masks*

Theia (https://www.theia-land.fr) delivers three masks (located in the MASK directory) for each image that should be used to eliminate saturated pixels (*_SAT.tif), clouds and cloud shadows pixels (*_NUA.tif) as well as pixels relating to water, snow, shadows or located outside the image (*_DIV.tif).

Step	QGIS handling
1. Theia masks merging	In **QGIS**: • Open the three associated masks for the year 2015: o LSAT_2015_SAT.tif o LSAT_2015_NUA.tif o LSAT_2015_DIV.tif In **Raster Calculator**: • Enter the following expression: ("LSAT_2015_SAT@1" + "LSAT_2015_NUA@1" + "LSAT_2015_DIV@1") = 0 • Save the file as MASK_2015.tif. In the same way, calculate the mask of the next year (LSAT_2016_SAT/NUA/DIV.tif), and name the output file: MASK_2016.tif.

Table 5.5. *Theia masks merging*

5.3.3.2. *Creation of the forest mask*

The forest mask is made here from the 2012 CORINE Land Cover layer. For France, it is advisable to use more accurate databases such as BD Topo® IGN (vegetation layer) or BD Forêt® IGN.

Step	QGIS handling
1. Preparation of the vector layer	In **QGIS**: • Open the vector layer of CORINE Land Cover: CLC12_RBFC_RGF.shp. The objective is to retain only the classes related to the forest; these are identified in the attribute field "CODE_12" by the codes 311, 312 and 313.

	In QGIS menu: • Click on the icon [icon] (Select features using an expression); • Enter the following expression: "CODE_12" = '311' OR "CODE_12" = '312'OR "CODE_12" = '313' • Save selection as: CLC_2012_FORET.SHP.

2. Search for bounding box coordinates of the study area	The objective is to convert CLC_2012_FORET.SHP into a raster that has the same dimensions (bounding box, number of pixels) as the study area raster.
	
	In the **Layers Panel**:
	
	• Double-click on MASK_2016.tif, the window of layer properties is displayed.
	
	In the **layer properties** window:
	
	• Click on tab **Metadata** [icon]
	   o In the **Properties** field:
	     ■ Move to **Layer Extent** and copy the coordinates.
	
	[icon] Metadata Layer Extent (layer original source projection)
	700140.0000000000000000,6611310.0000000000000000 : 810150.0000000000000000,6721320.0000000000000000
	
	• Paste the coordinates in a notepad for example and:
	   o Replace the characters ',' (comma) and ':' (colon) with a space;
	   o Insert the '-te' command at the beginning of the line;
	   o The final instruction is written as follows:
	
	-te 700140.0 6611310.0 810150.0 6721320.0

3. Raster conversion	In the **menu bar**: • Click on **Raster > Conversion > Rasterize (Vector to Raster)...** In the **Rasterize** window: • Choose the **Input file (shapefile)**: CLC_2012_FORET.SHP; • Save the output file as MASK_FORET.tif; • Click on the button **Edit** [icon] to modify the Gdal command line; • Copy/paste the notepad line just after gdal_rasterize;

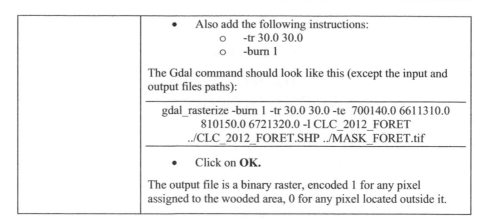

	• Also add the following instructions: o -tr 30.0 30.0 o -burn 1 The Gdal command should look like this (except the input and output files paths): gdal_rasterize -burn 1 -tr 30.0 30.0 -te 700140.0 6611310.0 810150.0 6721320.0 -l CLC_2012_FORET ../CLC_2012_FORET.SHP ../MASK_FORET.tif • Click on **OK.** The output file is a binary raster, encoded 1 for any pixel assigned to the wooded area, 0 for any pixel located outside it.

Table 5.6. *Creation of the forest mask*

5.3.3.3. *Masks merging*

The masks, once created, are merged before being applied to the changes image.

Step	QGIS handling
1. Masks merging	In the **menu bar**: • Click on **Raster > Raster Calculator...** • Enter the following expression: "MASK_2015@1" * "MASK_2016@1" * "MASK_FORET@1" • Name the output file: MASK_FINAL.tif.

Table 5.7. *Creation of the final mask*

5.3.3.4. *Applying masks to the changes image*

Once the merged mask is created, it is superimposed on to the changes image.

Step	QGIS handling
1. Input data verification	In **QGIS**: • Check for the following files: o DIFF_NDVI_2015-2016.tif o MASK_FINAL.tif
2. Mask integration	In the **menu bar**: • Click on **Raster > Raster Calculator...** • Enter the following expression:

	"MASK_FINAL@1" * "DIFF_NDVI_2015-2016@1" + ("MASK_FINAL@1" = 0) * -3
	• Name the output file: DIFF_NDVI_2015-2016_MASK.tif.
3. Specification of no data value	At this point, the image is ready. However, the –3 value assigned to invalid pixels is not saved in the image file header, which distorts the image statistics. In the **menu bar**: • Click on **Raster > Projections > Warp (Reproject)…** In the **Warp** window: • Choose the **Input file** DIFF_NDVI_2015-2016_MASK.tif; • Name the **Output file** DIFF_NDVI_2015-2016_MASK_NODATA.tif; • Check the **No data values** option o Enter the value -3; • Click on **OK**. 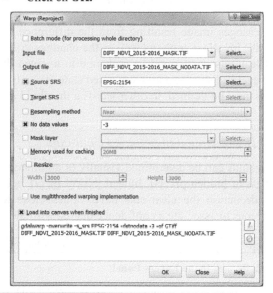
4. Verification of statistics	In the **Layers Panel**: • Double-click on DIFF_NDVI_2015-2016_MASK_NODATA.tif, the window of layer properties is displayed. In the **layer properties** window:

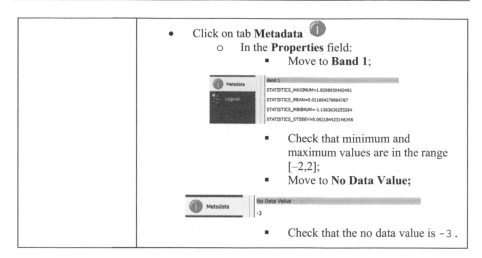

	• Click on tab **Metadata**
	○ In the **Properties** field:
	▪ Move to **Band 1**;
	▪ Check that minimum and maximum values are in the range [−2,2];
	▪ Move to **No Data Value**;
	▪ Check that the no data value is -3.

Table 5.8. *Applying masks to the changes image*

5.3.4. *Step 4: clear-cuts detection*

The changes image resulting from the difference in NDVI makes it possible to identify the invariant zones as well as the changes affecting land cover between two dates. Clear-cuts correspond to a change in a given direction, that is, the passage from a vegetated state to a non-vegetated state. To extract this information, the range of values must be defined, in other words a threshold must be applied to the changes image.

5.3.4.1. *Discretization of detected clear-cuts*

Thresholds can be determined visually or automatically based on image statistics (regardless of hidden pixels). Indeed, NDVI of $D1$ being subtracted from NDVI of $D2$, by deduction the clear-cuts have values close to the minimum. It may be advantageous to associate a confidence degree with the clear-cuts detection.

Step	QGIS handling

1. Search for changes image statistics	In the **Layers Panel**:
	
	• Double-click on raster DIFF_NDVI_2015-2016_MASK_ NODATA.tif, the window of layer properties is displayed.
	
	In the **layer properties** window:
	
	• Click on tab **Metadata** .
	    ○ In the **Properties** field:
	       ▪ Move to **Band 1**;

	▪ Note the values of mean m and standard deviation σ. Metadata — Legend Band 1 STATISTICS_MAXIMUM=1.8208930492401 STATISTICS_MEAN=0.011604279664767 STATISTICS_MINIMUM=-1.1363636255264 STATISTICS_STDDEV=0.062184423146348
2. Clear-cuts classification	Retrieve statistics (mean and standard deviation) of changes image. The confidence degree attributed to the detected clear-cuts is calculated as follows: • Clear-cuts with a low confidence degree $degré\ 1 = [(m - 2 * \sigma) \leq \Delta_{NDVI} < (m - \sigma)]$. • Clear-cuts with a medium confidence degree $degré\ 2 = [(m - 3 * \sigma) \leq \Delta_{NDVI} < (m - 2 * \sigma)]$. • Clear-cuts with a high confidence degree $degré\ 3 = [-2 \leq \Delta_{NDVI} < (m - 3 * \sigma)]$.
3. Identification of clear-cuts with a confidence degree	In the **menu bar**: • Click on **Raster > Raster Calculator…** • Enter the following expression: ("DIFF_NDVI_2015-2016_MASK_NODATA@1" >= -2 AND "DIFF_NDVI_2015-2016_MASK_NODATA@1" < (m – 3*σ)) * 3 + ("DIFF_NDVI_2015-2016_MASK_NODATA@1" >= (m – 3*σ) AND "DIFF_NDVI_2015-2016_MASK_NODATA@1" < (m – 2*σ)) * 2 + ("DIFF_NDVI_2015-2016_MASK_NODATA@1" >= (m – 2*σ) AND "DIFF_NDVI_2015-2016_MASK_NODATA@1" < (m – σ)) *Replace m and σ with the values found in the metadata.* • Name the output file: CLEAR_CUTS.tif. The result is a raster whose pixel values are between 0 and 3: • 0 = absence of clear-cuts; • 1 = clear-cuts with a low confidence degree; • 2 = clear-cuts with a medium confidence degree; • 3 = clear-cuts with a high confidence degree.

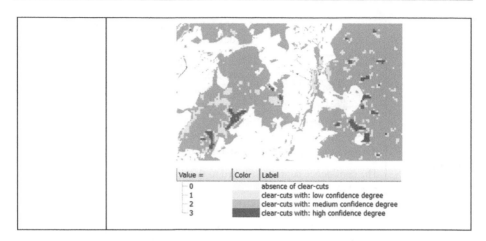

Value =	Color	Label
0		absence of clear-cuts
1		clear-cuts with: low confidence degree
2		clear-cuts with: medium confidence degree
3		clear-cuts with: high confidence degree

Table 5.9. *Classification of clear-cuts. For a color version of the table, see www.iste.co.uk/baghdadi/qgis2.zip*

5.3.4.2. *Filtering*

At the end of the thresholding, the raster created has coded values 1, 2 or 3, which identify the clear-cuts. However, these data comprise groups of pixels whose surface area is below the specifications, or even isolated pixels. A filter is applied to eliminate these elements and reduce the number of entities when exporting in vector mode.

Step	QGIS handling
1. Post-processings	In the **menu bar**: • Click on **Raster > Analysis > Sieve...** In the **Sieve** window: • Choose the **Input file**, for example: CLEAR_CUTS.tif; • Name the **Output file**, for example: CLEAR_CUTS_SIEVE.tif; • Enter a **threshold**: 11; *NB:* To be defined according to the image resolution and the minimum size sought for the clear-cuts, in this example the resolution is 30 m and the minimum size is 1 ha, that is to say nearly 11 pixels. • Enter the number of **Pixel connections**: 4.

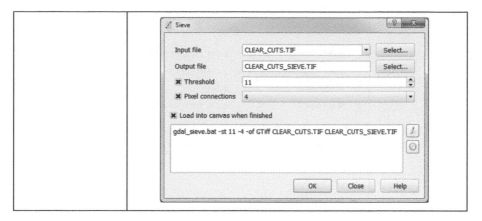

Table 5.10. *Raster filtering*

5.3.5. *Step 5: vector conversion*

The file of detected clear-cuts is finally exported to a vector format. The attribute table is updated to add the polygons area and possibly the false detections due to the presence of clouds on the satellite images.

5.3.5.1. *Vector export*

Converting raster to vector uses the GDAL library, installed by default on QGIS.

Step	QGIS handling
1. Exporting clear-cuts in vector	In the **menu bar**: • Click on **Raster > Conversion > Polygonize (Raster to Vector)…** In the **Polygonize** window: • Choose the **Input file (raster)**: CLEAR_CUTS_SIEVE.tif; • Name the **Output file**: V_CLEAR_CUTS.SHP; • Check the **Field name** option: DEGREE; • Check the **Use mask** option and select: CLEAR_CUTS_SIEVE.tif; *NB:* This option does not vectorize null pixels. • Click on **OK.**

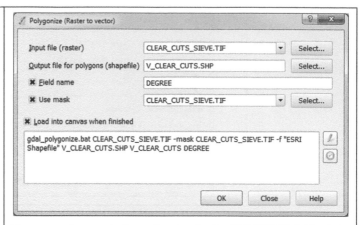

The output layer contains polygons. In the "DEGREE" field of the attribute table, clear-cuts are encoded in integers from 1 to 3.

2. Visual check	Open the satellite image LSAT_2016.tif and the previously created shapefile (V_CLEAR_CUTS.SHP).

Visually check the concordance between polygons and clear-cuts observed in the image.

Example of clear-cuts detection between summers 2015 (on the left) and 2016 (on the right):

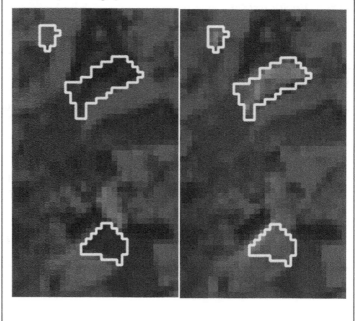

	Examples of false detections due to the presence of unmasked clouds: 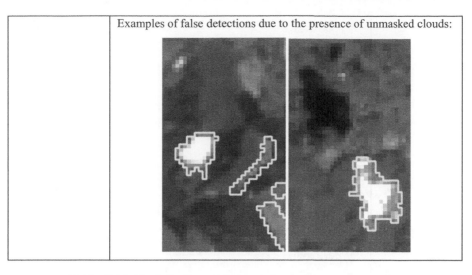

Table 5.11. *Vectorization and visual check. For a color version of the table, see www.iste.co.uk/baghdadi/qgis2.zip*

5.3.5.2. *Attribute table update*

The attribute table is enriched with a new field, the area (in hectare), so that the vector layer can be filtered and analysis can be carried out.

Step	QGIS handling
1. Calculation of polygons surface	Open the attribute table of vector file V_CLEAR_CUTS.SHP In the **attribute table**, • Open the field calculator by clicking on ▦. In the **field calculator**: • Check **Create a new field** option. • Enter the **Ouput field name**, for example: SURFACE. • Choose the **Ouput field type**: Decimal number (real). • Enter **Precision** = 2 (number of digits behind the decimal point). • In the **functions list**, move to **Geometry** and double-click on **$area.** • The surface area is calculated by default in the measure unit of the projection system, here the meter. In the field **Expression**, add /10000 to obtain a result in hectares. • Click on **OK.**

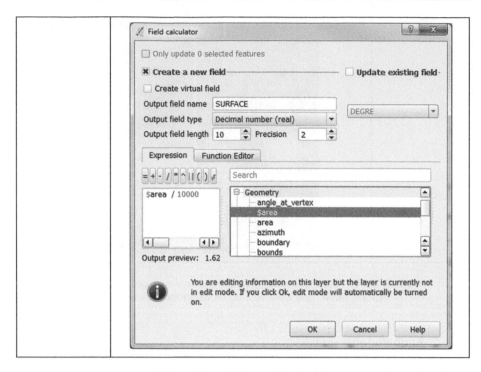

Table 5.12. *Attribute table update*

5.4. Bibliography

[OSE 16] Ose K., Corpetti T., Demagistri L., "Multispectral satellite image processing", in Baghdadi N., Zribi M. (eds), *Optical Remote Sensing of Land Surface: Techniques and Methods*, ISTE Press, London and Elsevier, Oxford, 2016.

Vegetation Cartography from Sentinel-1 Radar Images

6.1. Definition and context

Remote sensing is particularly well suited to monitoring terrestrial surfaces, especially vegetation. The cartography from spaceborne images allows the estimation of land use over large areas. In the case of forests, remote sensing data are widely used for the monitoring of deforestation due to lumbering, the monitoring of plantations and their certification or for applications in agro-forestry [PIK 02; OSU 15; NUS 11]. The cartography from spaceborne images is particularly used in the frame of REDD+ project[1] as shown by one widely used method, that is [PED 12] or [GEB 14].

The Sentinel-1 mission consists of the orbiting of two satellites (Sentinel-1A [S1A] and 1B [S1B]) carrying a Synthetic Aperture Radar (SAR) sensor. It allows, for the first time, free access to radar data on the ESA's (European Space Agency) scientific hub (https://scihub.copernicus.eu/dhus/#/home). Over land surfaces, the

Chapter written by Pierre-Louis FRISON and Cédric LARDEUX.

1 REDD+ (Reducing Emission from Deforestation and forest Degradation) is a policy set up in the post-Kyoto context. It was initiated by the tropical countries, responsible for a large proportion of global forests, during the United Nations Framework Convention on Climate Change (UNFCCC), in Montreal, Canada, in 2005. It aims to reduce carbon emissions into the atmosphere in order to combat global warming. Its basic principle is that countries able to reduce their emissions due to deforestation and forest degradation from a Business As Usual (BAU) benchmark level are compensated financially through a fund or a carbon market.

acquisitions are mainly realized in the *Interferometric Wide Swath* mode, characterized by a spatial resolution of about 20 m, a swath of 300 km, with two polarizations VV and VH. The orbital period of each satellite is 12 days. Consequently, acquisitions have been available since 2015 for time periods of 6 or 12 days depending on the study region. Over Europe, acquisitions are almost systematic, resulting in a wealth of information for temporal monitoring unmatched until then. The result is the need for massive data processing, the automation of which is essential in order to fully exploit acquisitions.

After a brief review of the main classification methods used in remote sensing, this chapter details the pre-processing of Sentinel-1 radar data (at Ground Range Detected (GRD) format) required for use in classification algorithms. These classifications aim to map main land use entities. The processing presented is developed in Python language, based on the Orfeo Toolbox (OTB) algorithms, and integrated into QGIS software, in order to make them accessible to non-specialists in radar data. The processing is grouped under the heading Sentinel-1 IW Batch Processing in the scripts of the QGIS Toolbox (in the Processing tab). An already configured version of QGIS (version 2.18 Las Palmas for Windows 7, 8 and 10), including OTB and the pre-processing programs detailed in this chapter, is available at: teledetection-radar.u-pem.fr/Book_ISTE_Press/software/QGIS_RemoteSensing_ 64bits.exe[2].

At the end of the installation, it is necessary to agree to install (if they are not already installed) the Visual Studio additional libraries proposed by default. Make sure that the C:\QGIS_RemoteSensing directory containing the software has been created. Its location in another directory can cause further problems. Figure 6.1 shows a color-composition made from a Sentinel-1 image generated by the processing detailed in this chapter. The programs developed for this purpose are integrated in the QGIS version and appear in the *Scripts, Sentinel-1 IW Batch Processing* section of the *Processing Toolbox* tab.

It should be noted that since April 2017, a Python script based on the OTB, which also allows the processing of Sentinel-1 data series, is available at tully.ups-tlse.fr/koleckt/s1tiling.

This script, which is very efficient in terms of computing time, is particularly suited to projects that require very regular land-use monitoring and over large territories. Due to time constraints, this script could not be implemented in the *QGIS_RemoteSensing* kit but will be implemented in future versions.

---

2 For Windows 32 bits systems, download GGIS_RemoteSensing_32bits.exe.

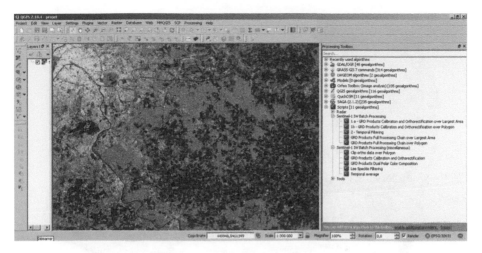

Figure 6.1. *Screenshot of a color-composite image displayed on the QGIS software produced from the Sentinel-1 acquisition over the Ile-de-France region on May 5, 2015. The programs detailed in this chapter appear under the headings "Sentinel- 1 IW Batch Processing" and "Sentinel-1 IW Batch Processing (miscellaneous)" in "Script" at the bottom of the "Processing Toolbox" tab. For a color version of the figure, see www.iste.co.uk/baghdadi/qgis2.zip*

6.2. Classification of remote sensing images

A classification algorithm aims to generate, from one or more remote sensing images, an image containing different thematic classes (see Figure 6.2). These classes are defined from the attribute properties of each of the pixels. The attributes are, for example, the reflectivity values in different polarizations or wavelengths, or the values of different textural parameters taking into account the neighborhood information of the pixels, etc. There are two types of classification: supervised and unsupervised.

Automatic classifications do not use any additional data for remote sensing images. Among the most widely used methods is the K-means algorithm. It is an iterative algorithm, with as an initial step, an arbitrary division of the attribute space into as many sub-regions as desired classes. Then, in step $N + 1$, it reassigns the pixels to the class of iteration N whose barycenter is the closest. This algorithm stops when the migrations of the class centers converge for a number of iterations less than a maximum number previously given by the user. They provide an overview of potential outcomes. Finer results can be obtained with supervised classifications described below.

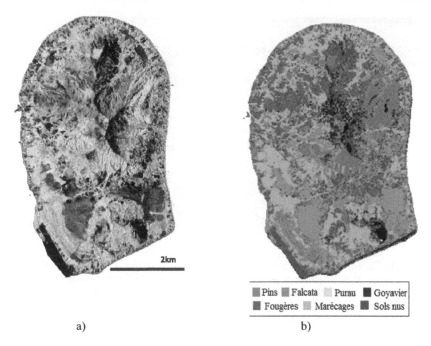

Pins ■ Falcata   Purau ■ Goyavier
■ Fougères   Marécages ■ Sols nus

a)   b)

Figure 6.2. *a) AIRSAR multi-polarized radar image (R: HH, V: VH, B: VV) acquired at L-band (λ = 24 cm) over the island of Tubuaï in French Polynesia. b) Results of an SVM (wide margin support) classification with seven different vegetation types [LAR 09]. For a color version of the figure, see www.iste.co.uk/baghdadi/qgis2.zip*

Supervised classifications require prior knowledge of all or part of the study area (from field surveys, or photo-interpretation of very high spatial resolution images). The operator will thus be able to define on the image *a priori* classes, called learning classes. This makes it possible to analyze the statistical properties of each of the classes during a phase called the learning phase, and allows the generation of a classification model. Then, in a second step, during the prediction phase, the algorithm will apply this model to the whole image. Among the most widely used supervised classification algorithms, the maximum likelihood can be cited, as can Support Vector Machines (SVMs), and Random Forest (RF), which have appeared these last few years. These last two methods are discriminative methods, making no *a priori* hypothesis on the distribution of the pixels, unlike generative methods. Moreover, they are specially adapted when one of the concerned images contains a large number of attributes (>10).

For a given pixel p (characterized by N attributes) to be assigned to a class among k, the maximum likelihood assigns to the pixel p the class C_i for which the conditional probability $P(C_i / p)$ $(1 \leq i \leq k)$ is maximum [CAN 10]. This method is often used when the attributes are not too numerous, because it is fast and gives good results. It assumes that the attributes of the pixels follow a normal distribution, which is not always the case.

SVM methods delimit the classes in the attribute (or feature) space by estimating the hyperplanes separating them best, that is the hyperplanes maximizing the distance (or margin) between two classes [BUR 98]. The projection of the feature space in a space of larger dimension is commonly performed in order to find optimal hyperplanes for the discrimination task. This projection comes down to the use of a kernel function. The most commonly used kernel function is the Gaussian function because of its ability to adapt to a wide variety of featured distributions. A prospecting of the whole feature space allows us to optimize two fundamental parameters: the standard deviation σ of the Gaussian function and the tolerated error, or cost parameter, called C, for the set of attributes.

The RF classification algorithm is based on the creation of a large number of decision trees, the whole constituting a forest [BRE 01]. Each decision tree is automatically created from a random sampling of the training points as well as the attributes. Each decision tree assigns to each of the pixels a given class. The final class that is chosen is the majority.

6.3. Sentinel-1 data processing

The pre-processing required for the use of Sentinel-1 data is summarized in Figure 6.3. It includes the following steps:

- radiometric calibration of data;

- ortho-rectification of calibrated data;

- application of filters to reduce the speckle effect;

- creation of color compositions (optional).

The first three processing steps allow Sentinel-1 data to be ready for use in the classification algorithms. The creation of colored compositions is generated automatically during the filtering of the speckle and therefore allows a color visualization of the filtered data. However, visual interpretation is more effective on unfiltered data. This is why this processing step, despite it being optional, is proposed.

Uncompressed downloaded
Sentinel-1 images
at GRD format

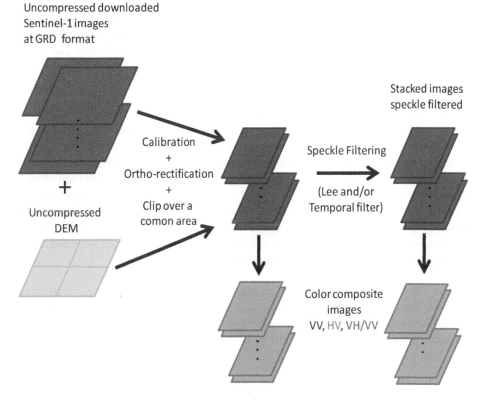

Figure 6.3. *Sentinel-1 GRD data processing chain. For a color version of the figure, see www.iste.co.uk/baghdadi/qgis2.zip*

6.3.1. *Radiometric calibration*

Radiometric calibration of radar images consists of converting the digital number of the image pixels into values directly related to the radar backscatter. Three different parameters can be derived: backscattering radar coefficients σ^0, γ^0, and radar brightness β^0 [POL 97].

6.3.2. *Ortho-rectification of calibrated data*

This step involves the transformation from a datum in image geometry to a datum in cartographic geometry, in other words to a projected image in a given cartographic system. The ortho-rectification of a radar image is carried out here by correlation of the image with a DEM (Digital Elevation Model) according to the method described in [SMA 11].

6.3.3. *Clip over a common area*

This step involves generating images of identical size (same number of rows and columns, same pixel size) on the study area. A pixel of a given latitude and longitude will therefore be referenced to the same row and the same column in each of the images.

6.3.4. *Filtering to reduce the speckle effect*

Two filters are implemented. They can be combined or applied independently of one another: the first is a filter operating only in the space domain [LEE 09]; the second operates, in addition, in the temporal dimension [QUE 01]. The latter is especially recommended in the case of Sentinel-1 data because the multitude of data acquired at different dates makes it possible to reduce the speckle particularly efficiently. This filter is defined according to:

$$ J_k = \frac{\langle I_k \rangle}{M} \sum_{i=1}^{M} \frac{I_i}{\langle I_i \rangle} \qquad\qquad [6.1] $$

M is the number of acquired images (with M different dates), J_k denotes the intensity of the pixel of the kth filtered image ($1 \leq k \leq M$), I_i denotes the intensity of the ith image to be filtered and $\langle I_l \rangle (1 \leq l \leq M)$ denotes the mean of the intensities performed over a local spatial neighborhood of the ith image. To this spatial dimension, performed over a neighborhood of fixed size, is added the temporal dimension appearing in the sum on the M images.

The Lee filter is an adaptive filter allowing us to reduce the speckle (in the spatial dimension only) over the homogeneous areas, while preserving the details (in other words, the heterogeneous areas) over which the size of the neighborhood decreases. Over homogeneous areas, it performs an average over the whole of the local neighborhood specified by the user. Over non-homogeneous areas, the size of the local neighborhood is as reduced as the heterogeneity is large. The homogeneous and heterogeneous areas are discriminated by comparing the coefficient of variation $c_V = \frac{\sigma}{\mu}$, σ and μ denoting, respectively, the standard deviation and the mean estimated on the actual local neighborhood. Over a homogeneous area, it is shown that the latter is constant and equal to $1/\sqrt{L}$, L being the number of views of the image, that is the number of statistically independent samples used to generate the image from the raw SLC (Single Look Complex) image. This value is the minimum value that can be observed in the image, since c_v will be greater than $1/\sqrt{L}$ as the considered area is heterogeneous. The number N indicated by the user for the number of views of the image is a parameter influencing the effects of the filtering, in the same way as the size of the local neighborhood. If $N > L$ ($1/\sqrt{L} < 1/\sqrt{1}/\sqrt{L} < 1/\sqrt{N}$), the effect of the filtering will be reduced (the homogeneous zones will be considered as heterogeneous and the size of the neighborhood will be reduced), and vice versa.

Figure 6.4 shows a comparison between two color composite images (see section 6.5) obtained from unfiltered and filtered GRD data.

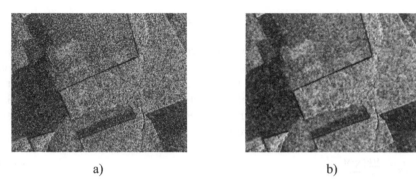

a)   b)

Figure 6.4. *Color composite images (R: VV; G: VH; B:VH/VV) generated from an original GRD datum a) and filtered with a Lee filter on a neighborhood of size 5×5 b). Sentinel-1 acquisition in IW mode on January 19, 2016 over the Juara region, Brazil. For a color version of the figure, see www.iste.co.uk/baghdadi/qgis2.zip*

6.3.5. *Generation of color compositions based on different polarizations*

The generation of color compositions facilitates the visual interpretation of the images. These compositions display the polarizations, VV, VH and their ratio VH/VV[3], respectively, in the red, green and blue channels on a logarithmic scale. Figure 6.5 shows color compositions produced from a RADARSAT-2[4] acquisition in polarization VV, VH and HH on Kourou in French Guiana. A colored composition in the VV, VH, VH/VV configuration (Figure 6.5(a)) with a similar color composition involving the HH polarization (Figure 6.5(c)) instead of VV (which can also be acquired by Sentinel-1) in the red, green, blue channels, respectively, is also shown. These two representations are to be compared with a

---

3 The polarization ratio (VH/VV) concern data in linear scale. When the data are in a logarithmic scale (decibels), this ratio becomes the difference (VH − VV)

4 Unlike Sentinel-1, the SAR sensor onboard RADARSAT-2 is fully polarimetric. Therefore, it allows the simultaneous acquisition of the same scene with the three polarizations HH, HV and VV.

representation with a colored composition involving the three polarizations (HH, VH and VV; Figure 6.5(b)) for which the interpretation is more evident. With three polarizations, the volume scattering (associated with dense vegetation) appears in green, the surface diffusion (bare soil or rough water surfaces) appears in blue and double bound mechanisms (urban areas facing radar, flooded vegetation, but also intervening on the bridge of Kourou, between the bridge and the river) appear in red.

In the case of dual polarization, whether in the VV + VH or HH + HV configuration, the dense vegetation appears in cyan-yellow, the bare soil surfaces appear in blue or red in the case of greater roughness and urban areas or flooded vegetation in bright orange. It can be noted that the areas of flooded vegetation, particularly the area parallel to the coastline to the north, are clearly detected with HH polarization, whereas they appear as low density vegetation with VV + HV configuration.

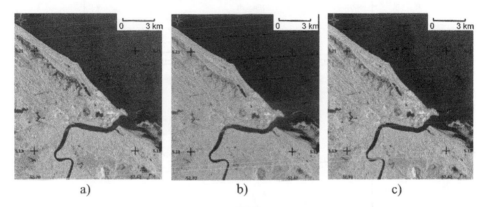

a)   b)   c)

Figure 6.5. *Colored compositions generated from data in dual polarization a) VV + VH; c: HH + VH), according to the convention red: VV (or HH); green: VH; blue: VH/VH (or VH/HH), and comparison with b) data acquired in the three polarizations (R: HH, V: VH, B: VV). RADARSAT-2 image acquired in Kourou, Guyana. For a color version of the figure, see www.iste.co.uk/baghdadi/qgis2.zip*

6.4. Implementation of the processing within QGIS

The various processes are accessible in QGIS from the previously configured version (integrating OTB and Python with the appropriate libraries). Its installation under Windows is detailed below.

The examples of data processing given in this chapter concern acquisitions of Sentinel-1 realized on January 19, September 27 and November 14, 2016, in the Juara region of Mato Grosso State, Brazil. These data (as well as the SRTM data required for georeferencing) are freely available on various space agency Websites, as detailed below. In addition, they are grouped together and can also be downloaded from the following Website: teledetection-radar.u-pem.fr/Book_ ISTE_Press/data/Juara/, within the directories *Sentinel-1* and *SRTM*, respectively.

Approach	QGIS manipulation
Installation of QGIS with OTB integration	Within the file manager, double click on program *QGIS_RemoteSensing_64bits.exe* that you have downloaded at the following address: teledetection-radar.u-pem.fr/Book_ISTE_Press/ software/QGIS_RemoteSensing_64bits.exe. • language of the installation wizard: English. 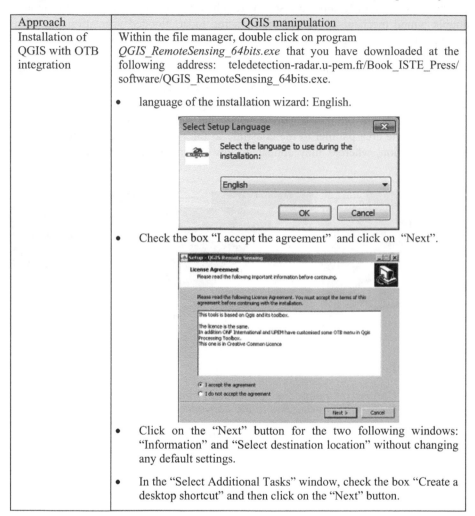 • Check the box "I accept the agreement" and click on "Next". • Click on the "Next" button for the two following windows: "Information" and "Select destination location" without changing any default settings. • In the "Select Additional Tasks" window, check the box "Create a desktop shortcut" and then click on the "Next" button.

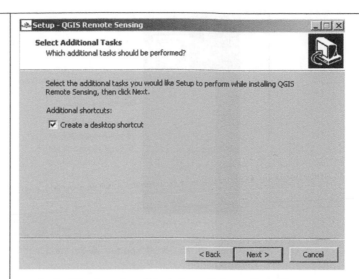

- In the "Ready to Install" window, click on "Install".

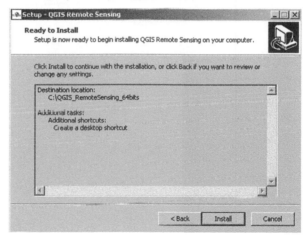

- Concerning the window "Completing the QGIS Remote Sensing Setup Wizard" window, there are two scenarios :

  – If QGIS has never been installed on your computer, run the default setup.

  – Otherwise, uncheck the three boxes checked by default.

- Click on the "Finish" button.

- Click on the "Next" button in the next window for Visual C++ installation.

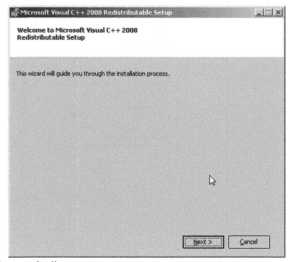

- Accept the license terms.

- Then click on "Install".

The installation is now finished.

To launch this version of the QGIS software, double click on the "QGIS_RemoteSensing" icon on the desktop

Make sure that this kit has been installed in the C:\QGIS_RemoteSensing_64bits directory. Installing it in another directory may cause errors later on.

Table 6.1. *Installation sequence of the configured version of QGIS software (integrating Orfeo Toolbox as well as Python libraries)*

6.4.1. *Downloading data*

The first step is to download the data of interest, in other words, the Sentinel-1 data and the DTM. These data must be stored in directories dedicated solely to them, one for the GRD data and the other for the DTM.

Approach	Manipulation
Downloading Sentinel-1 data	• In a Web browser, go to the ESA's scientific data portal: https://scihub.copernicus.eu. • Click on "OpenHUB". • Log in by clicking on the "LOGIN" tab at the top right of the page (for the first login, create an account requiring your email address and a password). • On the map, navigate/zoom to the region of interest and select the search criteria in the left tab. 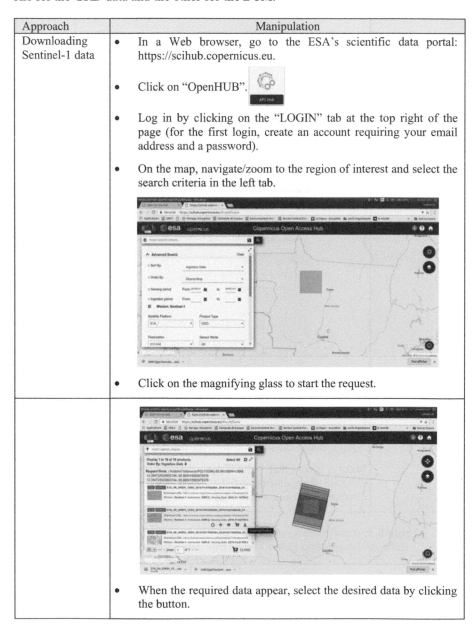 • Click on the magnifying glass to start the request.
	  • When the required data appear, select the desired data by clicking the button.

	• Store the Sentinel-1 GRD data in a directory dedicated only to these data.  • Uncompress the data. The directory will then contain as many sub-directories (each ending with the .SAFE extension) as dates of acquisitions. 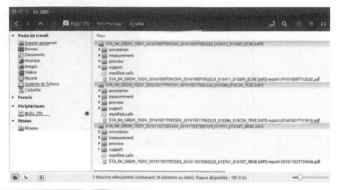
Downloading the SRTM DEM	• In a Web browser, go to the USGS Website: https://earthexplorer. usgs.gov. • Click on the "Login" tab at the top right of the page (for the first login, create an account requiring your email address and password). • On the world image, navigate/zoom on the area of interest. Delineate this area by clicking to mark the vertices of the associated polygon.

- On the left tab, click on the "Data Sets" button, then select "Digital Elevation", then SRTM 1 Arc-Second Global.

- Click on the "Results" button.

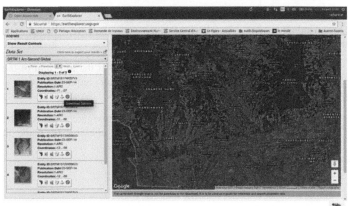

- You can view the footprints of each data by clicking on the button.

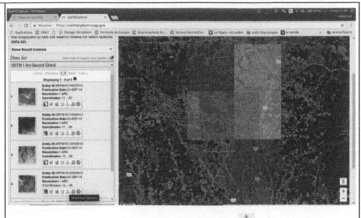

- Download the desired data by clicking on 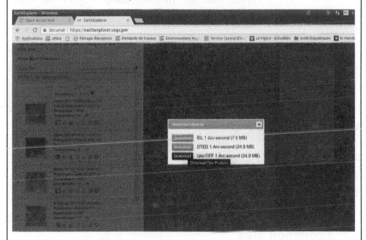.

- In the next window, select the TIFF format.

- Store all the SRTM datasets within a directory solely dedicated to them.

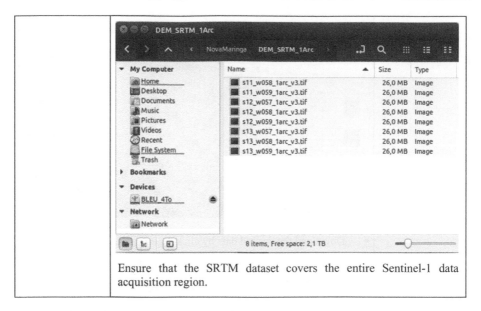

Ensure that the SRTM dataset covers the entire Sentinel-1 data acquisition region.

Table 6.2. *Data downloading (Sentinel-1 and SRTM). For a color version of the table, see www.iste.co.uk/baghdadi/qgis2.zip*

The processing codes that are used are written in Python and directly integrated in the version of QGIS configured with the OTB. They are located in the *C:\QGIS_RemoteSensing\qgisconfig\progessing\scripts* directory. They are accessible within QGIS in the *Processing Toolbox* tab, under *Script* and then *Sentinel-1 IW Batch Processing*.

6.4.2. Calibration, ortho-rectification and stacking of Sentinel-1 data over a common area

This step creates as many images as input images. The output images are all superimposed over a common area, with the same pixel size (and therefore the same number of rows and columns), and in the same European Petroleum Survey Group (EPSG) geographical projection. The programs used are accessible in the *Scripts*, *Sentinel-1 IW Batch Processing* section of the *Processing Toolbox* tab.

The menu *1a – GRD products Calibration and Orthorectification over Largest Area* allows this processing for the common area defined according to the greatest intersection of the dataset. The input directory (to be entered) is the directory containing the GRD data (hereinafter *rep0*). The backscatter parameter is to be chosen between *Sigma0* (by default), *Beta0* or *Gamma0*.

In the absence of a more accurate DEM, the required DEMs for the data ortho-rectification are those derived from 30 m resolution SRTM downloaded from the United States Geological Survey (USGS) Earth Explorer site: https://earthexplorer.usgs.gov. It is necessary to ensure that the DEM dataset covers the entire Sentinel-1 data acquisition region.

Data cropping requires us to define the same pixel size (*Output Pixel Size set*, at 10 m by default), and to choose output projection (EPSG) among those proposed by QGIS. Caution, the value given for the *Output Pixel Size* is in the units defined by the projection system. To obtain a size of 10 m, it is necessary to choose a projection system whose units are defined in meters (UTM type for example).

The output directory (to be entered, which will be called *rep1*) contains as many sub-directories as acquisition dates, in other words, the same architecture as the directory *rep0*. The processed images are saved in Geotiff format and are named *Sat_date_pol_par_Clip_Ortho.tif*, where *Sat* designates S1A or S1B, *date* is in format *yyyymmdd*, *pol* denotes the polarization ("VV" or "VH") and *par* corresponds to Sig0, Gam0 or Bet0 depending on whether the backscattering parameter is Sigma0, Beta0 or Gamma0. For example, the image S1A_20161009_VH_Sig0_Clip_Ortho.tif corresponds to the image acquired by Sentinel-1A, on October 9, 2016, in polarization VH, calibrated according to the parameter Sigma0, and ortho-rectified over a common area with the other images of the same parent directory.

The output directory also contains a color-composite image that corresponds to the time average for each pixel in a series of images (named *rep1/TempAverage_VV_VH_VII-VV_dB_firstdate_lastdate.vrt*, where *firstdate* and *lastdate*, in the format *yyyymmdd*, denote the first and last date in the time series), as well as the three .tif files starting with *TempAverage_* in the *VV*, *VH* and *VH – VV* associated polarizations. If the temporal series contains enough images, this results in an image whose speckle is almost no longer perceptible, having preserved the spatial resolution. This is to the detriment of temporal information that has completely disappeared. It may therefore be of interest to help delineate certain polygon characterizing classes in the subsequent classification step.

The menu *1b – GRD products Calibration and Orthorectification over Polygon* is analogous to the previous menu, but in this case the extraction area corresponds to that defined by a polygon (input polygon file). All other parameters to be entered are identical to those of *menu 1a*.

Approach	Manipulation
Running of processing : *1a – GRD products Calibration and*	• Verify that the Input Data Folder directory only contains the presence of the directories corresponding to each Sentinel-1 data acquisition.

Orthorectification over Largest Area	• Verify that the DEM folder only contains the DEM files (uncompressed, at Geotiff format). The *[optional] label* at the end of the sub-titles requesting the directory names comes from the OTB and cannot be deleted. It is not to be taken into account. The entry of the directory names is, on the contrary, mandatory in order to start processing. 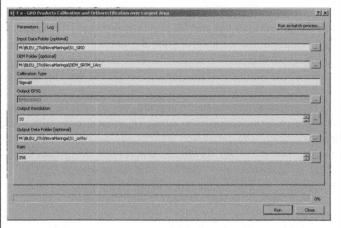 This may take a long time (the simple ortho-rectification of each image takes about 20 minutes on a 2.6 GHz 4-core computer with 16 GB of RAM). • The output directory has the following structure: 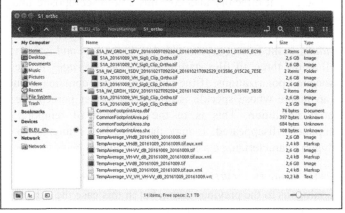

Table 6.3. *Menu 1a – GRD products Calibration and Orthorectification over Largest Area*

6.4.3. *Speckle filtering*

The data can then be processed according to the filter defined in equation [6.1] from menu *2 – Temporal Filtering*. The Spatial Neighborhood Scale corresponds to the *Spatial Window Size for Temporal Filter* parameter, which is set to 7 by default.

In order to enhance the filtering effect, we can also choose to first apply an adaptive Lee filter with the *Apply Lee PreFiltering* option, checked by default. In this case, the local neighborhood size is to be specified (*Spatial Window Size for Lee Filter*, 5 by default), as well as the number of views, *Looks Number for Lee Filter* (set to 5 by default, the number of views of GRD products).

The processed images are saved in Geotiff format with the same name as the original images, but with the suffix *_TempFilt_Wxtif*, where *x* is the parameter associated with *Spatial Window Size for Temporal Filter*.

By default, the processed images are in decibels (dB). However, they can be left on a natural scale by unchecking the "*Output in dB*" box.

In addition to the filtered images in the VV and VH polarizations, this processing also generates a VH/VV polarization ratio image. In addition, a QGIS virtual file is generated, allowing the three resulting images to be displayed as a colored composition (see section 6.3.5).

Approach	Manipulation
Speckle filtering	The *Input Data Folder* corresponds to the output directory of the previous step. It is necessary that all images in this directory contain the same number of rows and columns. The *Output Data Folder* to be entered will contain as many sub-directories as acquisition dates, in other words, the same architecture as the *rep1* directory.

The output directory has the following structure:

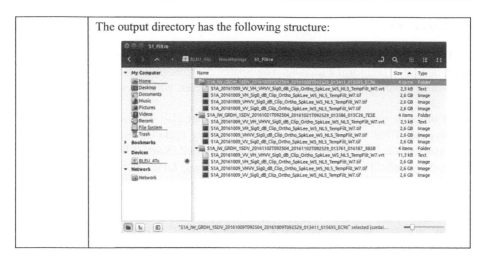

Table 6.4. *Speckle filtering*

After these two steps (sections 6.4.2 and 6.4.3), the processed images are ready to be used for classification algorithms.

A processing step grouping the steps described in sections 6.4.2 and 6.4.3 is also available in the *Sentinel-1 IW Bath Processing* menu. It is:

– *GRD Products Full Processing Chain over Largest Area*, grouping steps 1a and 2;

– *GRD Products Full Processing Chain over Polygon*, grouping steps 1b and 2.

In this case, the resulting data from processing 1a or 1b (see section 6.4.2) are not saved. Only data from step 2 (section 6.4.3) are saved in the output directory. They have the extension *Pol_Par_Ortho_Clip_TempFilt_Wx.tif*, *Pol* designating *VV* or *VH*, *Par* designating *Sig0*, *Gam0* or *Bet0* depending on whether the backscattering parameter is selected and *x* corresponding to the Spatial Window Size for the entered Temporal Filter parameter. Again, if the Lee pre-filtering has been selected, the extension will be *Pol_Par_Ortho_Clip_Tempfilt_Wx_pkLee_Wy_NLz.tif*, *y* and *z* corresponding to the Spatial Window Size for Lee Filter and the Looks Number for Lee Filter parameters, respectively.

6.4.4. *Other tools*

Different steps in the previous processing chains (sections 6.4.2 and 6.4.3) are also available in the case of specific requests. They are accessible in the *Sentinel-1*

IW Batch Processing menu (miscellaneous) and allow the operations detailed in this section.

6.4.4.1. *Clip of a set of ortho-rectified data over a given polygon*

This processing is available in the *Clip ortho data over Polygon* menu.

It makes it possible to extract a region, determined by a previously defined polygon, from data previously ortho-rectified (as, for example, by using programs 1a or 1b). The data to be entered are the same as for processing 1b (see section 6.4.2). The output directory (to be entered) contains as many sub-directories as acquisition dates, in other words, the same architecture as the input directory.

6.4.4.2. *Calibration and ortho-rectification over a dataset*

This processing is available in the *GRD products Calibration and Orthorectification* menu.

The difference with menu 1a or 1b is that the data, once calibrated and ortho-rectified, are not then extracted over a common region (corresponding to a given polygon or the greatest common intersection). The result files are saved in the output directory (of the same tree as the input directory) with the extensions *VH_Par_Ortho.tif* et *VV_Sig0_Par_Ortho.tif*, *Par* being the character string *Sig0*, *Gam0* or *Bet0* according to the selected backscatter parameter.

6.4.4.3. *Generation of colored compositions for a dataset*

This processing is available in the *GRD ProductsDual Polar Color Composition* menu.

It is particularly useful for creating colored compositions associated with unfiltered GRD data, which are easier to interpret visually than filtered data. It generates images in logarithmic scale (in decibels units) whose dynamics are better suited to human vision. The resulting files are saved in the output directory (of the same structure as the input directory). Each directory, associated with an acquisition date, contains the three files (Geotiff) required to create the colored composition, namely the VV, VH biased files and their VV/VH ratio. The names of the files in polarization VV and VH are identical to the original files, with the respective extensions *_dB.tif*. The file associated with the polarization ratio has the string: *_VHVV_* as well as the extension *_dB.tif*. The colored composition, that can be viewed within QGIS, corresponds to the virtual file ending by *dB.vrt*, and including the character string *VV_VH_VH-VV*.

Approach	Manipulation
Generation of color compositions	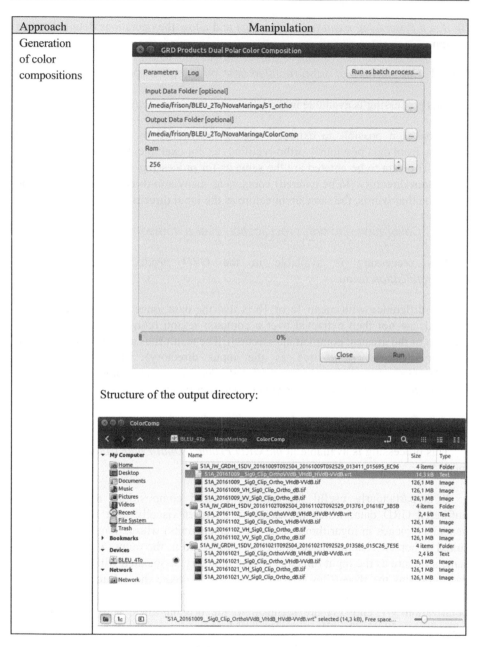 Structure of the output directory:

Table 6.5. *Color composition generation*

6.4.4.4. *Applying the Lee filter to a dataset*

This processing is available in the *Lee SpeckleFiltering* menu.

Here, only the Lee filter is applied, unlike the processing described in section 6.4.3, for which a temporal filtering of the data is systematically performed. The resulting files are saved in the output directory (of the same structure as the input directory) with the *SpkLee_Wx_NLy.tif* extension, x and y designating the spatial neighborhood size (7 by default) and the number of views (5 by default) that have been entered.

6.4.4.5. *Generation of a "Temporal Average" image*

This processing is available in the *Temporal Average* menu.

It calculates the temporal average of a data stack (superimposed and of the same sized images). Each pixel of the output image corresponds to the average of all the pixels at the same position of the images stack. In the same manner as in section 6.4.2, this program creates the time average for each polarization (VV, VH and VH-VV in a *.tif* file. It also creates a virtual file named ***rep0/TempAverage_VV_VH_VH-VV_dB_firstdate_lastdate.vrt*** (see section 6.4.2), where ***rep0*** is the input directory that was entered

6.5. Data classification

The application of the classification presented here is the (supervised) RF classification, which requires training polygons. They can be downloaded at the following address:

– tcledetection-radar.u-pem.fr/Book_ISTE_Press/data/Juara/vecteurs/polygones_entrainement.tar

Approach	QGIS manipulation
1. Creating an image containing all the desired bands	Load all the filtered Sentinel-1 images, corresponding to each acquisition date, at polarization VV, VH and their ratio VH/VV into QGIS. To do this, click the *Add Raster Layer* button ⬛ . In the *Open a GDAL Supported Raster Data Source* window, move to the S1_Filter directory, and then type * .*tif* in the top right box beside the magnifying glass. You will then see all the files (nine in total) with .*tif* extension in the sub-directories of the *ColorComp_Filt* directory.

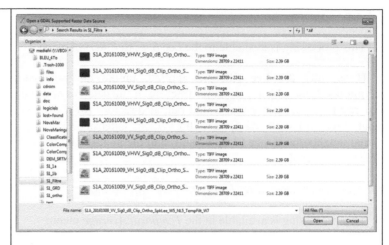

Select all the files.

- Images may appear black in QGIS because of a wrong default setting of their histogram. In this case, zoom in at full resolution of the image by clicking on the button 🔍 . Then, adjust the histogram of the image to the viewing area by clicking on 📊. (if the image remains black, zoom in and/or move to another area and repeat the operation). Then return to the entire image view by clicking 🔲.

- It remains to group these nine images into one single, virtual file, which is composed of nine bands, and will be used for the classification. To do it:

  - First, order the images by increasing date following the sequence VV; VH; VH/VV;

  - Check that the display box of each band is checked (i.e. each band is visible);

  - Then *Raster > Miscellaneous > Build Virtual Raster*, and save the virtual image in the Classification directory under the name *S1A_IW_3dates.vrt*.

<table>
<tr><td></td><td>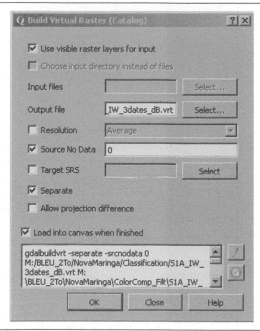</td></tr>
</table>

2. Sub-image extraction	In order to shorten the processing times, we will extract a sub-image. • Download the training polygons at the following address: *teledetection-radar.u-pem.fr/Book_ISTE_Press/data/Juara/vecteurs/ polygones, entrainement.tar*. • Move this file to the Classification directory, decompress it, and then load the *polygon_training.shp* vector layer. • Zoom into the entire *polygon_training layer.shp*. • Then, in the *Raster > Extraction - Clipper* menu, with the *Extent* box of *clipping mode* checked, drag and drop the mouse over the entire image displayed, and save it in the Classification directory under the name *extrait_S1A_IW_3dates.tif*.

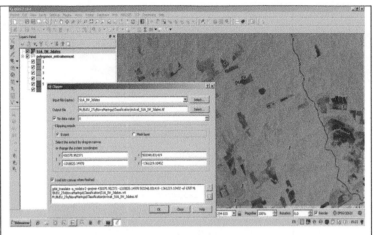

- Open the iage *extrait_S1A_IW_3dates.tif*.

The different stages of classification under OTB have been grouped in the sub-menu *1 — Classification* of the *Orfeo Toolbox* menu in the *Processing Toolbox* tab.

- In the *Processing Toolbox* tab, in the sub-menu *1 – Classification 1* of the *Orfeo Toolbox menu*, click *Compute Image Statistics* to calculate the statistics for the image. Select the *extract_S1A_IW_3dates* image as Input images and *Classification/Statistics_S1_3dates.xml* as the output file (Output XML file).

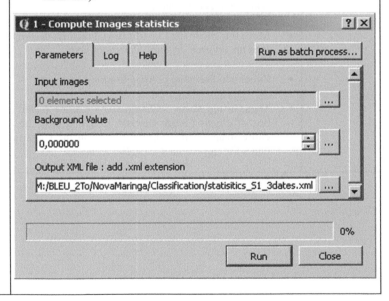

- For the learning phase of the Random Forest classification, click *2 – Train Random Forest Image Classifier* in the *1- Classification* menu of Orfeo Toolbox.

This learning phase uses a polygon file that must necessarily contain a field indicating the code of the class. This field must be encoded as an integer (and not text that is proposed by default). The name of the field must then be entered in the field *"Field Name containing ... "*.

Fill in the fields as shown in the following screenshot :

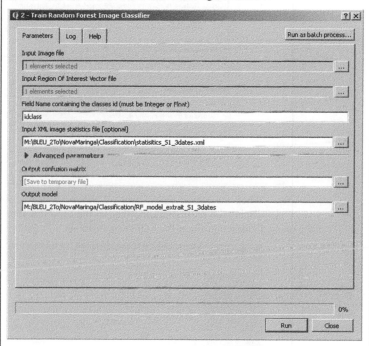

- Then *click on 3 – Create Image classification* in menu *1 – Orfeo Toolbox classification* and fill in the fields as shown in the following window.

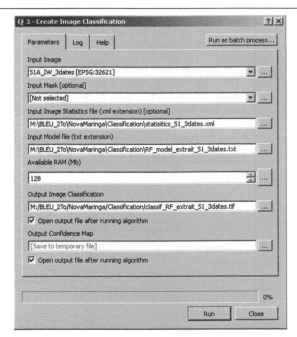

- When the *classif_RF_extrait_S1_3dates* image is opened in QGIS, display the properties, and then in the *Style* thumbnail, complete the fields as shown in the following window:

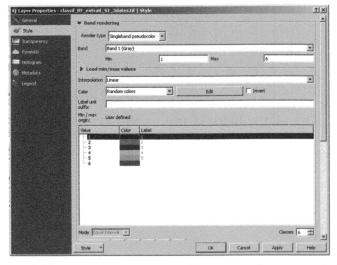

By clicking twice on each of the colors, change them to match the colors of the polygons in the training classes of the *polygones_entrainement* layer.

- In order to reduce the small groups of isolated pixels, apply the post-processing algorithm "Sieve" by going to *Raster -> Analysis -> Sieve* and filling in the fields as follows (this example shows a minimum size of 50 pixels, or 0.5 ha with a pixel datum at 10 m):

Sieve	? X
Input file	classif_RF_extrait_S1_3dates.tif ▼ Select...
Output file	assif_RF_extrait_S1_3dates_sieve.tif Select...
☑ Threshold	50
☑ Pixel connections	4
☑ Load into canvas when finished	

gdal_sieve.bat -st 50 -4 -of GTiff
M:/BLEU_2To/NovaMaringa/Classification/classif_RF_extrait_S1_3dates
.tif
M:/BLEU_2To/NovaMaringa/Classification/classif_RF_extrait_S1_3dates

OK   Close   Help

Figure 6.6 shows the result of the classification.

Table 6.6. *Stages for the classification of speckle-filtered Sentinel-1 data. For a color version of the table, see www.iste.co.uk/baghdadi/qgis2.zip*

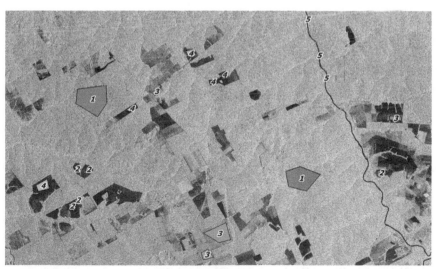

a)

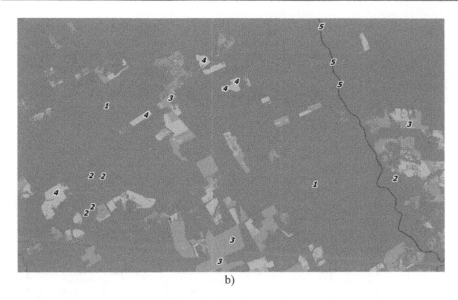

b)

Figure 6.6. *a) Color composition of a Sentinel-1 acquisition extract, acquired on January 19, 2016, over the Juara region, Brazil, on which are superimposed the training polygons of five classes (1: forest, 2: agricultural fields A, 3: agricultural fields B, 4 bare soils, 5: water. Two types of agricultural fields are distinguished due to different temporal behaviors). b) Result of the Random Forest classification from the three Sentinel-1 acquisitions. For a color version of the figure, see www.iste.co.uk/ baghdadi/qgis2.zip*

6.6. Bibliography

[BRE 01] BREIMAN L., "Random forests", *Machine Learning*, vol. 45, p. 5, 2001.

[BUR 98] BURGES C.J.C., "A tutorial on support vector machines for pattern recognition", *Data Mining and Knowledge Discovery*, vol. 2, pp. 121–167, 1998.

[CAN 10] CANTY M.J., *Image Analysis, Classifications, and Change Detection in Remote Sensing*, CRC Press, Taylor and Francis, 2010.

[GEB 14] GEBHARDT S., WHERMANN T., RUIZ M.A.M. *et al.*, "MAD-MEX: automatic wall-to-wall land cover monitoring for the Mexican REDD-MRV Program using all Landsat data", *Remote Sensing*, vol. 6, pp. 3923–3943, 2014.

[LAR 09] LARDEUX C., FRISON P.-L., TISON C. *et al.*, "Support vector machine for multifrequency SAR polarimetric data classification", *IEEE Transactions on Geoscience and Remote Sensing*, vol. 47, no. 12, pp. 4143–4152, 2009.

[LEE 09] LEE J.S., WEN J.H., AINSWORTH T.L. *et al.*, "Improved sigma filter for speckle filtering of SAR imagery", *IEEE Transactions on Geoscience and Remote Sensing*, vol. 47, no. 1, pp. 202–213, 2009.

[MAC 09] MacQueen J.B. "Some methods for classification and analysis of multivariate observations", *Proceedings of 5th Berkeley Symposium on Mathematical Statistics and Probability*, vol. 1, pp. 281–297, 2009.

[NUS 11] Nuscia Taibi A., Munoz N., Ballouche A. *et al.*, "Désertification en Zone Soudano-Sahélienne (pays Dogon, Mali)? Apport de la Cartographie et du Suivi Diachronique des Parcs Agroforestiers par Télédétection Satellitaire et Aérienne", *Proceedings of the 25th International Cartographic Conference*, Paris, 2011. Available at: https://hal.archives-ouvertes.fr/hal-01105201/document.

[OSU 15] O'Sullivan R., Estrada M., Durschinger L. *et al.*, "FCMC/Terra Global Capital and VCS Staff, Technical Guidance for Jurisdictional and NestedREDD+ Programs", *VCS Guidance*, 2015. Available at: http://database.v-c-s.org/sites/vcs.benfredaconsulting.com/files/JNR%20Guidance%20-%20Part%20B_%20final%202%20June%2015.pdf.

[PED 12] Pedroni L., "Verified carbon standard – a global benchmark for carbon – methodology for avoided unplanned deforestation VM0015 – v1.1", 2012. Available at: http://database.v-c-s.org/sites/vcs.benfredaconsulting.com/files/VM0015%20Methodology%20for%20Avoided%20Unplanned%20Deforestation%20v1.1.pdf.

[PIK 02] Piketty M.-G., Veiga J.B., Poccard-Chapuis R., *et al.*, "Le potentiel des systèmes agroforestiers sur les fronts pionniers d'Amazonie brésilienne", *Bois et Forêts des Tropiques*, vol. 2, pp. 75–87, 2002. Available at: bft.cirad.fr/cd/BFT_272_75-87.pdf.

[POL 97] Polidori L., *Cartographie Radar*, Gordon and Breach Science Publisher, Taylor and Francis, 1997.

[QUE 01] Quegan S., Yu J.L., "Filtering of multichannel SAR images", *IEEE Transactions on Geoscience and Remote Sensing*, vol. 39, no. 11, pp. 2373–2379, 2001.

[SMA 11] Small D., "Flattening gamma: radiometric terrain correction for SAR imagery", *IEEE Transactions on Geoscience and Remote Sensing*, vol. 49, no. 8, pp. 3081–3093, 2011.

Remote Sensing of Distinctive Vegetation in Guiana Amazonian Park

7.1. Context and definition

7.1.1. *Global context*

The Guiana Amazonian Park (Figure 7.1) has satellite imagery available all over its territory. The SPOT-5 sensor offers 10 m imagery resolution that allows for the detection of distinctive vegetation [GRA 94], distinguished from the tropical forest (which looks like broccoli florets from above). Remote sensing was used to detect those distinct sets of vegetations. The objective is to identify, with the help of their spectral signature, these distinctive species and map them over the national park area. The resulting map can contribute towards other French Guiana research projects such as forest landscape projects [GON 11] and forest habitat maps [GUI 15].

Eventually, this methodology can be used to monitor the evolution of these vegetations, including for climate change adaptation analysis [HOL 17], and their evolution in the medium and long term. Currently, there is little knowledge about the origin of these vegetations. Bamboo (*Guadua* spp.) lifetime should be around 30 years [NEL 05], but the small amount of data available does not allow us to follow their spatial evolution [OLI 07].

Chapter written by Nicolas KARASIAK and Pauline PERBET.

Figure 7.1. *Location of the French Guiana Amazonian Park. For a color version of the figure, see www.iste.co.uk/baghdadi/qgis2.zip*

7.1.2. *Species*

Jean-Jacques de Granville identified several vegetation species in the south of French Guiana [GRA 94]. Our study is concerned only by the species that are grouped within at least 1,000 m² (Figure 7.2) and thus not scattered vegetations.

– **Low vegetation** (cambrouses – bamboo or heliconia forests on the riverside; or Pripris – wetland vegetation on the riverside);

– **Mauritia flexuosa forest;**

– **Açaí palm forest** (*Euterpe oleracea*);

– **Rocky outcrop** (Inselberg with some vegetation);

– **Parinari forest** (found in the Waki river area) in djougoun-pété soil (pits of 1 m diameter and from 30 to 50 cm depth).

a)   b)   c)

Figure 7.2. *a) Açaí palm forest, b) Inselberg and c) bamboo forest (credits: Daniel Sabatier - IRD). For a color version of the figure, see www.iste.co.uk/baghdadi/qgis2.zip*

7.1.3. *Remote sensing images available*

There are different types of imagery available in French Guiana to realize this project (Table 7.1). Images from SPOT-5 were chosen for this work because they were the best resolution imagery available at the beginning of the project in 2014. However, SPOT-5 does not operate anymore, so the methodology was made compatible with Sentinel-2 and Landsat-8.

Because the project area is in an equatorial climate, it is impacted by clouds most of the year. To avoid missing areas, the methodology used a superposition of several images. At the end, 55 images, covering 23 SPOT-5 scenes, were used to cover the French Guiana Amazonian Park (Figure 7.3).

Landsat-8 and Sentinel-2 were assessed to continue this methodology with free and easy access of data. The benefit of Sentinel-2 is its high temporal resolution of 5 days. It is thus easier to obtain complete coverage of the area every year.

Sensor	Period	Spatial resolution		National Park coverage*
		Multispectral	Panchromatic	
Landsat (5-7-8)	Since 1990	30 m	15 m	100%
SPOT-4	1998–2013	20 m	-	100%
SPOT-5	2002–2015	10 m	2.5 m	100%
SPOT (6-7)	2014–today	6 m	1.5 m	70%
Pléiade	2012–today	2 m	0.5 m	5%
Sentinel-2**	2016–today	10 m	-	100%
Lidar	2014	1 m		Two mountains (Itoupe and Inini)
SAR Sentinel-1**	2016–today	10 m		100%

*Cover for all available imagery.
**Available after this work.

Table 7.1. *Satellite imagery available*

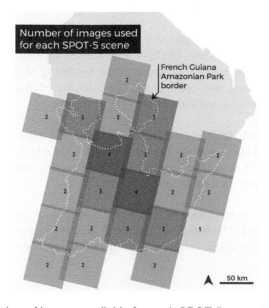

Figure 7.3. *Number of images available for each SPOT-5 scene to get a free cloud coverage. For a color version of the figure, see www.iste.co.uk/baghdadi/qgis2.zip*

7.1.4. *Software*

The entirety of this work was achieved with open-source software (Table 7.2). This software provides easier propagation and best replicability of the method. In order to easily enable replication of this project, it is important to use well-known classification algorithms. This is why dzetsaka plugin has been developed. This name means "objects used to see the world" (such as a camera or a satellite) in the Teko language (an Amerindian language spoken in Oyapock river, south east of French Guiana).

Software used (version)	Extension
Monteverdi	Optical calibration
QGIS (2.14.14)*	GDAL library Dzetsaka Grass (6.4.3) Orfeo Toolbox: - Band Math - ComputeConfusionMatrix

*Only the extensions to activate or not present by default are listed.

Table 7.2. *Software used for this project*

7.1.5. *Method implementation*

Throughout the method presentation, an example with SPOT-5 imagery will be detailed.

SPOT-5 satellite imagery was acquired from the SEAS Guiana project. SEAS offers free French Guiana SPOT-5 archive images to the public organization located in this territory (see: http://www.guyane-sig.fr/?q=node/88).

The same method was validated for Sentinel-2 and Landsat-8. Sentinel-2 provides spatial resolution close to SPOT-5, but with a better temporal resolution (every 10 days). The work initiated with SPOT-5 can continue with these new sensors.

Procedure and tool location can slightly differ according to the operating system or the QGIS version used. This chapter has been verified with the use of Windows 10, QGIS 2.14.14, Grass 6.4.3 and OTB 5.0.0.

7.2. Software installation

To carry out this work, the dzetsaka plugin should be installed from the QGIS extension list. However, dependencies should be installed first: Python-pip, Scikit-learn, Scipy, Grass 6.4.3, Orfeo Toolbox 5 and Monteverdi.

The dzetsaka plugin was designed to classify images directly from QGIS. To unlock all its capabilities, several dependencies should be installed first to use all the famous classification algorithms (like Random Forest (RF) and Support Vector Machine (SVM)). For Windows installation, it is better to use OsGeo.

7.2.1. Dependencies installation available in OsGeo

To update or install QGIS with all the dependencies, run the OsGeo installer (available on the download page at Qgis.org), then select advanced installation. Several libraries are available, but dzetsaka only needs Python-pip and SciPy. However, to do atmospheric correction, Monteverdi is recommended.

> **Setup dependencies with OsGeo:**
> *- Advanced installation > Select Packages > Check Python-pip, scipy, Monteverdi, grass 6, otb-bin*

To install a dependency, click on the Libs category and select Install instead of Default after having entered the dependency name in the search box.

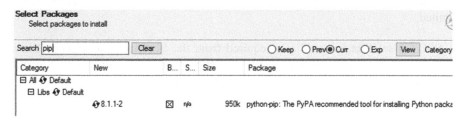

Figure 7.4. *Dependencies installation in OsGeo*

Once dependencies have been installed, open QGIS and click on Menu > Processing > Options > Providers, then activate GRASS commands.

Grass default directories are:

– C:\OSGeo4W64\apps\grass\grass-6.4.3 for Grass 6.4.3;

– C:\OSGeo4W64\apps\msys for MSYS, essential to link GRASS and QGIS.

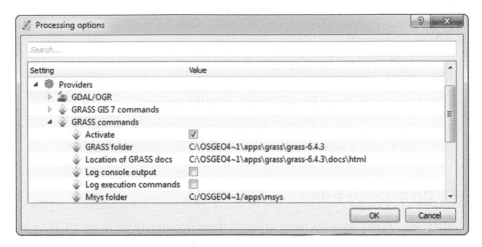

Figure 7.5. *Settings for Processing Toolbox*

If needed, Orfeo Toolbox default directories are:

– C:\OSGeo4W64\bin for tool directory;

– C:\OSGeo4W64\apps\orfeotoolbox\applications for applications directory.

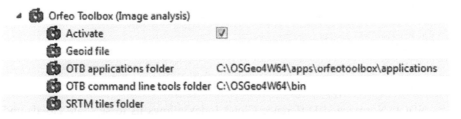

Figure 7.6. *Orfeo ToolBox settings in QGIS*

7.2.2. *Installation of scikit-learn*

Scikit-learn is the reference Python library for automatic learning. It is necessary to use famous classification algorithms such as RF, SVM or K-Nearest Neighbors (KNN). Unavailable in the OsGeo tool, a little manipulation in the OsGeo terminal to implement this feature is required.

> **Setup scikit-learn library:**
> *- Windows Start Menu > Open OsGeo4W Shell in administrator mode...*

In the windows Start Menu, open OsGeo4W Shell (in administrator mode), and execute this command:

```
pip install scikit-learn
```

With Linux, if pip is not installed, just type in the terminal:

```
sudo apt-get install Python-pip
sudo pip install scikit-learn
```

7.2.3. Dzetsaka installation

Like any QGIS plugin, in order to install dzetsaka, simply go to the QGIS open menu/Plugin/Install and search for dzetsaka. Click on install and the plugin will automatically download and install itself.

Since the introduction of dzetsaka version 2.1, the plugin is now compatible with the processing toolbox, which means you can use dzetsaka in a processing chain.

> **QGIS features:**
> - Setup dzetsaka: *Plugin > Setup or manage plugin > Search for dzetsaka*

7.3. Method

Lexicon:

– NDCI: Normalized Difference Cloud Index allows us to separate clouds and shadows from the rest of the image.

– TOA: Top-of-Atmosphere corresponds to the reflectance value at the top of atmosphere.

– ROI: Region of Interest, these are the polygons or points containing data used to classify an image.

Figure 7.7 shows the processing of the six main steps for forest tree mapping:

1) convert images in TOA reflectance;

2) clouds mask creation with the NDCI;

3) ROIs creation;

4) classification;

5) correction of results (remove noises with a sieve);

6) final processing step and merge all results into one.

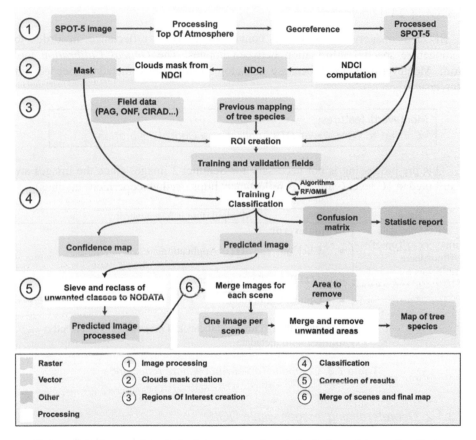

Figure 7.7. *Processing chain for mapping forests in Guiana Amazonian Park. For a color version of the figure, see www.iste.co.uk/baghdadi/qgis2.zip*

7.3.1. *Image processing*

Image processing (Figure 7.8) involves taking the raw SPOT-5 image and converting it to TOA reflectance. Then, if some spatial lag is visible, georeferencing the TOA image.

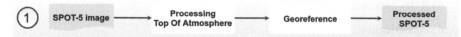

Figure 7.8. *Step 1 diagram: convert image in TOA reflectance. For a color version of the figure, see www.iste.co.uk/baghdadi/qgis2.zip*

7.3.1.1. *TOA reflectance*

SPOT-5 imagery needs to be converted to TOA reflectance in order to homogenize spectral information between images. The *Optical Calibration* tool from Monteverdi is used because it automatically finds metadata needed for conversion.

> **Monteverdi features:**
> - Convert to TOA: *View > OTB Application > Optical Calibration*

This pre-processing is not necessary for Sentinel-2 images since the images are distributed in 1C level (with TOA reflectance: https://scihub.copernicus.eu/dhus).

Steps	Manipulation in Monteverdi
Convert raw data imagery to Top-of-Atmosphere reflectance	In **Monteverdi:** • Click on: View/OTB Application. In the **OTB Application**, open: • Optical Calibration. Load the SPOT-5 image and check if parameters are correctly fulfilled with metadata.

Table 7.3. *Compute Top-of-Atmosphere reflectance*

7.3.1.2. *Georeferencing images within QGIS*

Images must be correctly georeferenced to limit spatial lag. If they are not sufficiently georeferenced, the use of a georeferenced dataset is needed to correct spatial differences. In our case, we use the IGN (National Institute of Geographic and Forest Information) SPOT-5 mosaic of French Guiana.

> **QGIS features:**
> - Plugin activation: *Plugin > Manage and install plugins > Select Gdal Georeferencer*
> - Georeferencing an image: *Raster > Georeferencer > Georeferencer ...*

Steps	QGIS steps

Georeferencing an image	**In QGIS:**
	
	• Load a georeferenced reference image;
	
	• Open the georeferencer tool:
	Raster > Georeferencer > Georeferencer
	
	In Georeferencer tool:
	
	• File > Open Raster
	
	• Settings > Transformation Settings:
	
	    ○ Select the same Coordinates Reference System as the reference raster;
	    ○ Choose the transformation type Polynomial (1 or 2).
	
	• After having added at least three points, activate the linking between the georeferencer and QGIS;
	
	    ○ View > Link Georeferencer to QGIS;
	    ○ This will create the same spatial extent between your reference image and the future georeferenced image.
	
	• The average error appears at the bottom right, in the edge of the window. The closer the error is to 0, the better your georeferencing.

Table 7.4. *Georeferencing images*

7.3.2. *Cloud mask creation*

Step 2 (Figure 7.9) explains how to create a cloud mask for each image.

Figure 7.9. *Step 2 diagram: cloud mask computation. For a color version of the figure, see www.iste.co.uk/baghdadi/qgis2.zip*

7.3.2.1. *Index computation (NDCI) for cloud mask creation*

The climatic context of French Guiana leads to high cloud coverage, including in the dry season (from August to November). It is common to have a cloud coverage rate higher than 50%. Creating a cloud mask is fundamental.

NDCI is used to filter clouds and shadows. It also takes into account the difference between the short wave infrared (SWIR, band 4 of the SPOT-5 image) and the red (band 2):

$$NDCI = \frac{Red-SWIR}{Red+SWIR} \qquad\qquad [7.1]$$

The *Band Math* tool available in Orfeo Toolbox will be used to compute this index. The tool does not support the NDCI function but it does support NDVI. Simply indicate band 2 and 4 to obtain the NDCI (red band and SWIR), or 2 and 3 for NDVI (red band and near-infrared), but the function will remain incorrectly named "NDVI".

QGIS features:
- Compute NDCI: *Processing Toolbox > Orfeo Toolbox > Miscellaneous > Band Math...*

Steps	QGIS steps
Compute NDCI	In **Processing Toolbox:** • Orfeo Toolbox > Miscellaneous > Band Math In **Band Math**: • Write the formula: "ndvi(im1b2,im1b4)" Save the result with the NDCI suffix (example: SPOT5_NDCI.tif).

Table 7.5. *Calculation of the NDCI*

7.3.2.2. *Cloud mask*

NDCI values scale from −1 to 1. In order to have a quality cloud mask, it is necessary to manually check the threshold above which the clouds are totally filtered. As a general rule, the values that allow us to filter the clouds oscillate between 0.40 and 0.45. For each image, this threshold must be correctly defined and visually validated.

The formula used in the *Band Math* tool is:

$$im1b1 <= 0,40 ? 0 : 1 \tag{7.2}$$

This formula replaces pixels that have a value higher than 0.4 by 1, if not the value will be 0.

> **QGIS features:**
> - Mask creation: *Processing toolbox > Orfeo Toolbox > Miscellaneous > Band Math...*

Sentinel-2 and Landsat-8 images are provided with cloud and shadow masks. However, in French Guiana these masks miss lot of clouds and shadows. The mask made from the NDCI was also reproduced for these sensors.

The mask is then directly used when classifying with the dzetsaka plugin, automatically detecting the _mask suffix in addition to the filename.

Steps	QGIS steps
Mask creation	In **Processing Toolbox:** • Orfeo Toolbox > Miscellaneous > Band Math In **Band Math:** • Write the formula: "im1b1 <= 0.4 ? 0:1" Verify the threshold (here 0,40) is enough to remove clouds and shadows. Then save the results with _mask suffix (example: SPOT5_mask.tif).

Table 7.6. *Creation of the cloud mask*

7.4. Processing

7.4.1. *Creating training plots*

Step 3 (Figure 7.10) consists of creating for each image a shapefile layer containing the ROI available in this image (polygons that contain data used for classifying the image, also known as training plots).

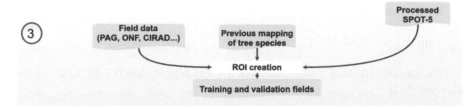

Figure 7.10. *Step 3 diagram (creating ROI). For a color version of the figure, see www.iste.co.uk/baghdadi/qgis2.zip*

Class	Type
1	Parinari
2	Low vegetation
3	Mauritia flexuosa forest
4	Inselberg
5	Açai forest
6	Clouds
7	Shadows
8	Light clouds
9	Water
10	Standard canopy
11	Nonstandard canopy

Table 7.7. *Nomenclature of the attribute table of a training plot (in green, the classes of vegetations to map). For a color version of the table, see www.iste.co.uk/baghdadi/qgis2.zip*

These polygons are identified with field data or by photo-interpretation. Identification of different classes must be saved as a whole number. For optimal quality, creation of ROIs needs to be precise. Some rules must be respected:

– ROIs must select pixels having a homogenous spectral signature and avoid borders with vegetations mixes.

– There must be a sufficient number of pixels in each of the classes studied.

– It is advisable to select areas distributed throughout the image in order to avoid errors induced by the biogeographic effect.

To improve knowledge of the study area, several helicopter flights were carried out to collect field reference data.

> **QGIS features:**
> - Vector creation: *Layer > Create Layer > New Shapefile Layer...*

Steps	QGIS steps
1. Vector creation	In **QGIS** : Create a new vector file: • Layer > Create Layer > New Shapefile Layer; • Choose Polygon type; • Choose the same coordinate reference system (CRS) as your image; • Create a new field called *class* "whole number", and a field *Type* "Text data" as shown in Table 7.7; • Save (example: SPOT5_ROI.shp);
2. Create new ROI	• Toggle editing by clicking on the editing icon ; • Create polygons of vegetations, clouds and shadows (remember to use the nomenclature label for the *class* column).

Table 7.8. *Creation of the ROIs*

7.4.2. *Classification with dzetsaka plugin*

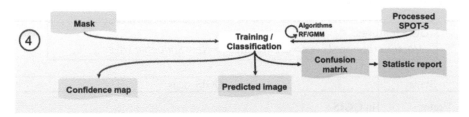

Figure 7.11. *Step 4 diagram (classification). For a color version of the figure, see www.iste.co.uk/baghdadi/qgis2.zip*

7.4.2.1. *List of algorithms available*

7.4.2.1.1. Random Forest (RF)

The RF algorithm is an automatic learning method based on the creation of different decision trees. This algorithm has proven itself in several works, including tree species classification [FAS 16].

In Figure 7.12, three decision trees are shown: two trees ($x2 < 5$ and $x3 < 5$) allowing the prediction of class 2. x1 is the pixel value in band 1, x2 in band 2 and x3 in band 3. One of the trees is based on pixel value in the infrared ($x2 < 5$ for example), which considers whether the pixel value is lower than 5 on this band, in which case the algorithm will predict it as water, or whether the tree will be more detailed in order to separate the species.

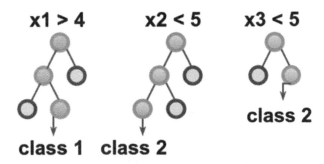

Figure 7.12. *How Random Forest works. For a color version of the figure, see www.iste.co.uk/baghdadi/qgis2.zip*

7.4.2.1.2. Support Vector Machine (SVM)

SVM is a renowned algorithm that is based on several kernels (linear, polynomial, Gaussian, etc.) in order to better adapt to the problem to be solved. The kernel implemented in dzetsaka is the Radial Basis Function (a Gaussian kernel) that provides high quality results in tree species classification [SHE 16].

Support vectors (circles touching dotted lines in Figure 7.13) are samples that are nearest to the separation margin (red line) and those that will have more importance for the classification. Circles in green and blue represent two different classes. x1 and x2 are their values in two different spectral bands (red and infrared for example).

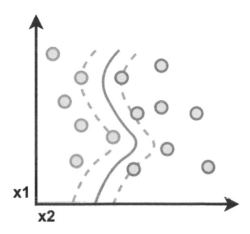

Figure 7.13. *How SVM works. For a color version of the figure, see www.iste.co.uk/baghdadi/qgis2.zip*

7.4.2.1.3. K-Nearest Neighbors (KNN)

The KNN algorithm is one of the oldest. After having defined the number of neighbors, the algorithm will class the pixel according to the value of the nearest neighbor's samples. This algorithm is very fast and needs only one parameter: the number of neighbors. In dzetsaka, this parameter is chosen in a cross-validation method to maximize quality. In Figure 7.14, circles in blue and green represent two classes. x1 and x2 are values in spectral bands (red and infrared for example). We can observe that if the number of neighbors is set to 3 (K = 3), or 6 (K = 6), the predicted class of the yellow circle will stay in the green class. To be precise, with three neighbors, two are from the green class and one is from the blue.

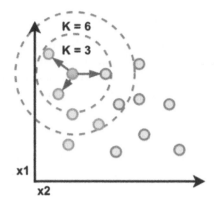

Figure 7.14. *How K-Nearest Neighbors works. For a color version of the figure, see www.iste.co.uk/baghdadi/qgis2.zip*

7.4.2.1.4. Gaussian Mixture Models (GMM)

GMM is an algorithm that is based on the sum of several Gaussian ones in order to classify the pixel. The closer the pixel is (in its reflectance values), the more important it will be. This algorithm is the fastest available in dzetsaka and is natively available (you do not need to install several dependencies). However, GMM is not suitable for hyperspectral or multitemporal images (it needs to be manually configure in the code), so use it with caution. We recommend using it only on multispectral images.

7.4.2.1.5. Which algorithm to use?

There is no algorithm that is better than another. It depends on the work carried out and on the data to be classified. There are several indices to evaluate the quality of an algorithm, including Cohen's kappa coefficient and Global Accuracy. The higher these indices are, the better the classification is. Although the SVM algorithm is generally the most efficient, it is not suitable for large ROIs (more than 10,000 pixels). RF is a good alternative, but it could be interesting to use other algorithms such as GMM because of its quick execution.

7.4.2.2. *Use of dzetsaka*

7.4.2.2.1. Making a classification

Dzetsaka plugin allows the user to classify images with different algorithms. The first thing to do is to link ROIs with the same projection system of the input image.

Once the classification is done with dzetsaka, a raster is generated and each pixel has the label of the predicted class. It is also possible to save the algorithm model (i.e. rules of the classification, to predict this model with another image), the confusion matrix or a confidence map (each pixel will have the confidence for the predicted class and data are scaled between 0 and 1; 1 means total confidence, 0 no confidence.)

> **QGIS features:**
> - Image classification: *dzetsaka interface > Perform the classification...*

Steps	QGIS steps
1. Image classification	In the **dzetsaka plugin:** • Select image to classify; • Select shapefile containing ROI; • Select field containing typology (*class* column); • In the optional tab, select the mask if the name doesn't end with yourImageName_**mask**.tif; • Click on Settings button to choose the algorithm ⚙; 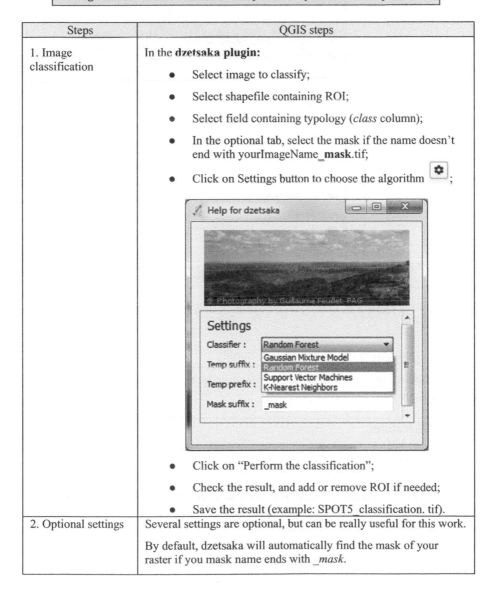 • Click on "Perform the classification"; • Check the result, and add or remove ROI if needed; • Save the result (example: SPOT5_classification. tif).
2. Optional settings	Several settings are optional, but can be really useful for this work. By default, dzetsaka will automatically find the mask of your raster if you mask name ends with _*mask*.

In the optional tab, it is possible to save the confusion matrix (*Save matrix*) that will be its CSV extension, or save the confidence map.

When saving the matrix, dzetsaka will put the split (pixels to keep for validation and pixels to keep for training) to 50%, which means half the pixel will be used for training. You can change this setting.

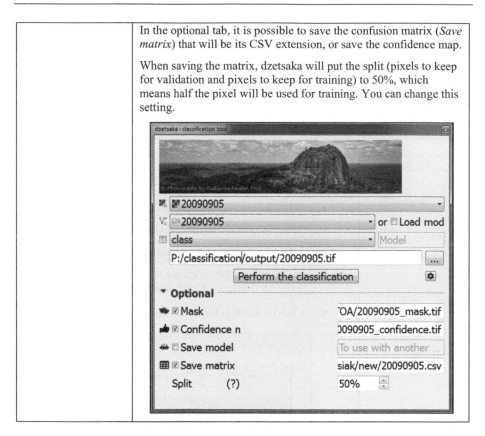

Table 7.9. *Making a classification with dzetsaka plugin*

7.4.2.2.2. Assessing the classification quality

Two tools allow us to assess the classification quality. The first one enables us to show the algorithm confidence per pixel.

> **QGIS features:**
> - Classification Quality assessment: *Check the case confidence map box in dzetsaka*

To obtain the confidence map, you need to check the box before classification. The result is a raster with values scaled from 0 to 1, 1 being maximum confidence and 0, minimum. It can be used to show spatially where the algorithm lacks information and where the confusion is higher. It allows us to add ROIs, or to show on the ground why the algorithm has no confidence in these cases.

Steps	QGIS steps
Obtain the confidence map	In the **dzetsaka** plugin: • In the interface, check the confidence map box; • Once classification is done, a raster will be generated: The lower the value (in black here), the more the algorithm is not confident.

Table 7.10. *Making a confidence map for assessing classification quality*

QGIS features:
- Assessment of the classification quality: *Dzetsaka > Confusion matrix...*

The second method is the creation of a confusion matrix. Once the confusion matrix box is checked, the split (percentage of pixels kept beside the classification) will be automatically set to 50%, which means 50% of the pixels will be used for classification, and 50% will be used to validate the classification. It allows the algorithms to class the image with only half of the pixels.

To obtain the global accuracy (percentage of pixels that are successfully classified) or the kappa coefficient (global accuracy minus the hazard effect), the dzetsaka menu needs to be opened, followed by the confusion matrix. This tool requires your ROI, the column that contains your labels and the classified raster. Once run is initiated, you will obtain your kappa and global accuracy score.

Steps	QGIS steps
Get Kappa coefficient and global accuracy.	**In QGIS:** • Click on: Dzetsaka > Confusion matrix **In the Confusion Matrix windows:** • Load your classification raster; • Load your shapefile; • Select the field containing typology (*class* column); • Click on Compare. In our example, kappa is 89.14% (0.89), and global accuracy is 95.17%, which means 95.17% of predicted pixels have the same label as those that are trained. 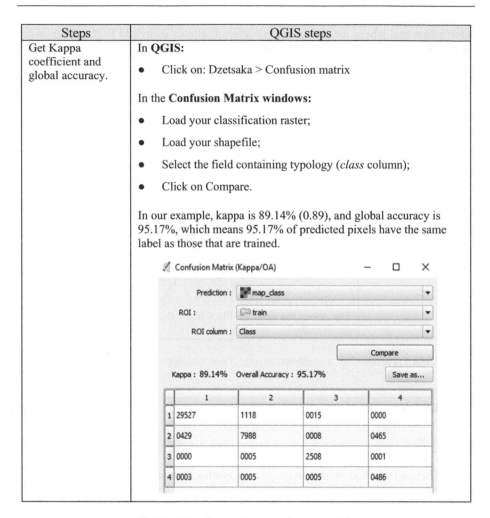

Table 7.11. *Compute a confusion matrix*

7.4.3. *Post-classification*

Once each classification is done, we can observe some noise. To remove this noise, a sieve must be applied to the raster, which means objects below a certain size (in pixels or in area) will be changed to this majority neighbor. Step 5 (Figure 7.15) corrects classification defaults and allows us to prepare the classification for the final step.

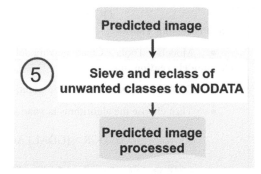

Figure 7.15. *Step 5 diagram (post-classification processing). For a color version of the figure, see www.iste.co.uk/baghdadi/qgis2.zip*

To facilitate the merging of several images, values of unwanted classes will be changed to "no data". In this case, clouds, shadows and water will be converted to the 0 value.

The *Band Math* tool allows the reclassification of classes 6, 7, 8 and 9 (corresponding to clouds, shadows, light clouds and water) to 0 with the following expression:

$$\text{im1b1} >= 6 \ \&\& \ \text{im1b1} <= 9 \ ? \ 0 : \text{im1b1} \tag{7.3}$$

To write an expression, a question that contains several arguments is required ("&&" for "and", and "||" for "or"). Once the question is proposed ("?"), the first answer is the true one (in our case "0"), then the answer if the condition is wrong (":"). In this case, if the condition is not satisfied (im1b1 is 3 for example), *Band Math* will assign its im1b1 value (so value "3").

In order not to carry out the same processing twice, we will create a Processing Toolbox in QGIS. This feature allows us to chain different functions and batch them. For example, it is possible to create a model that will create an NDCI image, then compute the cloud mask and classify the image. To introduce you to the model creation, a short model will be created to sieve the predicted image and to reclass unwanted classes.

QGIS features:
- Create a model: *Processing Toolbox > Models > Tools > Create new model...*
- Filter objects giving their size: *GDAL/OGR > [GDAL] Analysis > Sieve...*
- Reclass an image: *Orfeo Toolbox > Miscellaneous > Band Math*

Steps	QGIS steps
1. Create a model	In the **Processing Toolbox:** • Models > Tools > Create new model
2. Filter small objects	In the **model creation tool:** • Double click on a raster on the right; • Then choose the algorithms tab and search for: ° GDAL/OGR > [GDAL] Analysis > Sieve In the **Sieve window**: • Select the image (predicted image); • Set the **threshold to 10 pixels**, which means a sieve of 1,000 m² , which removes noises of single pixels and represents a suitable area to describe forests; • Set the **pixel connection to be 8**, which means considering all neighbors from this pixel. Connection 4 only takes into account neighboring pixels; • In sieved, write "*raw_sieve*"; • Click on OK.
3. Image reclass	In the **Model creation** tool: • Select the Band Math algorithm: ° Orfeo Toolbox > Miscellaneous > Band Math • In Input image, use the previously processed *raw_sieve* file; • Write the formula:

	"im1b1>= 6 && im1b1 <= 9 ? 0 : im1b1" • In field *OutputImage*, write the given name that will be requested for use: here *reclass_sieve*. 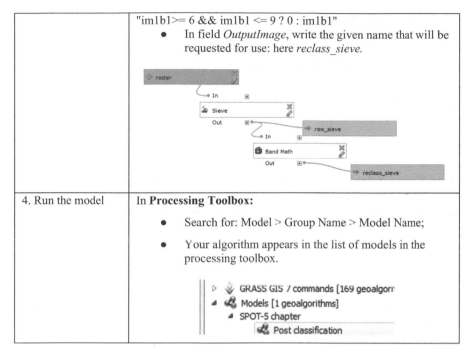

4. Run the model	In **Processing Toolbox:**
	
	 • Search for: Model > Group Name > Model Name;
	
	 • Your algorithm appears in the list of models in the processing toolbox.
	
	 ▷ 🌱 GRASS GIS / commands [169 geoalgorr
	 ⊿ 🐿 Models [1 geoalgorithms]
	  ⊿ SPOT-5 chapter
	   🐿 Post classification

Table 7.12. *Sieving and reclassifying the classification. For a color version of the table, see www.iste.co.uk/baghdadi/qgis2.zip*

7.5. Final processing

The last step (Figure 7.16) consists of merging every predicted image and deleting unwanted areas, such as habitat or mining territory.

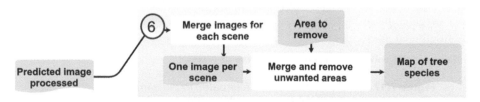

Figure 7.16. *Step 6 diagram (merging results and final processing step). For a color version of the figure, see www.iste.co.uk/baghdadi/qgis2.zip*

7.5.1. *Synthesis of predicted images*

The climate of tropical humid forests induces a high cloudiness in images. For greater accuracy mapping, it is necessary to overlay the classification results of different dates. Ideally, these images should be close in time to avoid disturbances (e.g. cuts). However, the climate condition in Guiana makes it difficult to obtain several images a year.

The *r.mapcalculator* tool from Grass 6 allows us to generate a classification synthesis for each scene, while removing pixels corresponding to the standard canopy forests. Grass is recommended for this step because it handles images of different sizes, which is very useful when the spatial extent is not the same between several acquisition dates. If you are working with similar spatial extents, we recommend using the Orfeo Toolbox classification merging tools: FusionOfClassification (majority vote, or Dempster–Shafer using the confusion matrix).

Inselbergs are best shown in the drying season (around October), so data from 2009 (A image in Figure 7.17) are the most important. However, there was a calibration problem with the satellite sensor on this date, so only the inselbergs could be extracted.

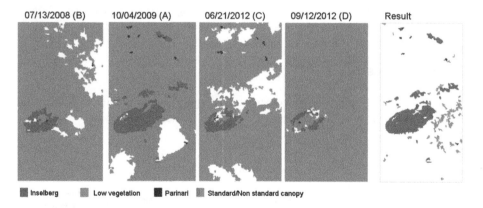

Figure 7.17. *Synthesis example of four dates in the same SPOT-5 scene. For a color version of the figure, see www.iste.co.uk/baghdadi/qgis2.zip*

On the 690-344 scene extract (Figure 7.17), the *r.mapcalculator* of Grass allows us to remove standard canopy and non-standard canopy (classes 10 and 11) in order to later merge different scenes between them.

Then a priority rule is given to favor the 2009 image (which is image A), except for class 2 (low vegetation). Be careful not to let empty space appear at the end of the expression or your script will generate an error.

```
if((isnull(A)  |||  A==0  |||  A==2  |||  A==10  |||
A==11),if((isnull(B)  |||  B==0  |||  B==10  |||
B==11),if((isnull(C)  |||  C==0  |||  C==10  |||
C==11),null(),C),B),A)
```

To translate this expression: if image A's pixel has a null value, or a value of 2, 10 or 11, we look to see if image B has a value different from null, 10 or 11, or we take image C (if different from null, 10 or 11). If no condition is satisfied (i.e. all pixels have the 10 value), the expression will be classified with a null value (white background shown in Figure 7.17). It is important to use the three bars ||| to represent the "or" because in Grass it also means using null values.

> **QGIS features:**
> - Data synthesis: *Processing Toolbox > Grass Commands > Raster > r.mapcalculator...*

Steps	QGIS steps

Select images and classes to prioritize	From the classification results, such as the confusion matrix or the confidence map, select an order of the classes and images to be prioritized.
	
	 In **Processing Toolbox**:
	
	 • GRASS commands > Raster > r.mapcalculator
	
	 In **r.mapcalculator** :
	
	 • Load images;
	
	 • Write the expression (A is image 1, B is image 2). Be careful not to let an empty space appear at the end of the expression.
	
	 • Example:
	 `if((isnull(A)|||A==0|||A==2|||A==10|||A==11),`
	 `if((isnull(B)|||B==0|||B==10|||B==11),if(`
	 `(isnull(C)|||C==0||| C==10 |||`
	 `C==11),null(),C),B),A)`

Table 7.13. *Synthesis for each scene*

7.5.2. *Global synthesis and cleaning unwanted areas*

The final result per scene is merged to obtain a unique map of the whole study zone. However, some anthropized areas have no interest in this mapping. In addition, deforestation zones are easily confused with inselberg during classification. A final cleaning that takes account of the land cover data is therefore necessary. A mask is created from agricultural and mining areas from the land cover layers.

QGIS features:
- Global synthesis: *Processing Toolbox > GDAL/OGR > Miscellaneous > Build Virtual Raster…*
- Merge vector layers: *Processing Toolbox > QGIS geoalgorithms > Vector general tools > Merge vector layers…*
- Buffer: *Processing Toolbox > QGIS geoalgorithms > Vector geometry tools > Variable distance buffer…*
- Vector rasterization: *Processing Toolbox > Orfeo Toolbox > Vector Data Manipulation > Rasterization (image)…*
- Remove unwanted areas: *Processing Toolbox > Orfeo Toolbox > Miscellaneous > Band Math…*

Steps	QGIS steps
1. Global synthesis	**In Processing Toolbox**: • GDAL/OGR > Miscellaneous > Build Virtual Raster **In Build Virtual Raster tool:** • Load all the scenes (each synthesis by scene); • Give a name to the virtual raster; • Unselect "Layer Stack"; • Click on OK.
2. Buffer	**In Processing Toolbox:** • QGIS geoalgorithms > Vector geometry tool > Fixed distance buffer **In Fixed distance buffer:** • If you are unsure of the areas to be removed, we recommend that you create a buffer. Choose the distance (100 for 100 m here because it is a metric projection). To check the measure unit, refer to the spatial reference site. For example, for Lambert-93 (EPSG: 2154): http://spatialreference.org/ref/epsg/2154/html); • Check the Dissolve *result* to merge buffers; • Save the result.

3. Vector rasterization	In **Processing Toolbox:** • Orfeo Toolbox > Vector Data Manipulation > Rasterization (image)… In **Rasterization (image)** tool: • Load vector areas to remove; • Load the global synthesis; • Set the background value to 1; • Select rasterization mode to binary; • Set the foreground value to 0; • Save the result.
4. Temporary final result	In the **QGIS** menu: • Raster > Raster Calculator… In **Raster Calculator** tool: • Load layer areasToRemove@1 (means band 1 from areasToRemove image); • Multiply it by the fusion_scenes image: " areasToRemove@1" * "fusion_scenes@1" • Save the result.
5. Final result	In **Processing Toolbox:** • GDAL/OGR > Conversion > Translate In **Translate** tool: • Input image: last result; • Null value: 0. In advanced parameters: • Raster type: byte; • GeoTIFF options: LZW; • Save the final result.

Table 7.14. *Global synthesis and cleaning*

7.5.3. *Statistical validation – limits*

Field data are very difficult to access in southern Guiana. Despite some helicopter and field missions, the data gathered led to too much imprecision. Indeed, from the helicopter, information has important spatial lag, whereas ground data potentially concern smaller formations, not detectable with a spatial resolution of 100 m².

To overcome this problem, a training game containing new training plots was created from the 2015 SPOT-6 mosaic. These are new ROIs that are visible by photo-interpretation from this image and it served as a validation set for the final result obtained previously. The mosaic is correctly georeferenced, and has the advantage of having a better spatial resolution.

A confusion matrix is obtained using the *ComputeConfusionMatrix* tool from Orfeo Toolbox. Inselbergs, low vegetation and Parinari forests have shown very good classification results. However, inselbergs and low vegetation classes should be detailed, as they include several species. Results for *Parinari* species and Açai palm forests are not convincing. These confusions appear to be related to the fact that palm areas are generally very small, and there are only a few geolocalized learning areas.

> **QGIS features:**
> - Compute the confusion matrix: *Processing Toolbox > Orfeo Toolbox > Learning > ComputeConfusionMatrix(vector)...*

Steps	QGIS steps
1. Create a reference layer	In the same way, as when creating training plots, digitize polygons from a new image (ideally with better spatial resolution), or use only reference field data.
2. Compute the confusion matrix	In the Processing ToolBox: • Orfeo Toolbox > Learning > ComputeConfusionMatrix (vector) In **ComputeConfusionMatrix**: • Input image (final classification); • Input reference vector data (ROI); • Set Field name (class); • Value for nodata: 0; • Save the result as csv type.

Table 7.15. *Validate the final classification*

7.6. Conclusion

This classification follows a standard method of landcover mapping with automatic analysis of remote sensing images. The contribution of this chapter is its implementation in an open-source and user-friendly tool: QGIS. Moreover, with the processing toolbox, it now allows us to combine multiple algorithms (and to use them in batch). This ability is essential for remote sensing works on large areas or with a large number of images, and this is the case in French Guiana.

7.7. Bibliography

[FAS 16] FASSNACHT F., LATIFI H., STEREŃCZAK K. *et al.*, "Review of studies on tree species classification from remotely sensed data", *Remote Sensing of Environment*, vol. 186, p. 64, 2016.

[GON 11] GOND V., FREYCON V., MOLINO J.F. *et al.*, "Broad-scale spatial pattern of forest landscape types in the Guiana Shield", *International Journal of Applied Earth Observation and Geoinformation*, vol. 13, p. 357, 2011.

[GRA 94] GRANVILLE J.-J., "Les formations végétales primaires de la zone intérieure de Guyane", *Forêt guyanaise: gestion de l'écosystème forestier et aménagement de l'espace régional*, SÉPANGUY, p. 244, 1994.

[GUI 15] GUITET S., BRUNAUX O., DE GRANVILLE J.J. *et al.*, Catalogue des habitats forestiers de Guyane, DEAL Guyane, available at http://www.onf.fr/lire_voir_ecouter /++oid++4cc4/@@display_media.html, 2015.

[HOL 17] HOLM J., KUEPPERS L., CHAMBERS J., "Novel tropical forests: response to global change", *New Phytologist*, vol. 213, no. 3, pp. 988–992, 2017.

[NEL 05] NELSON B.W., BIANCHINI M.C., "Complete life cycle of southwest Amazon bamboos (Guadua spp) detected with orbital optical sensors", *XII Simpósio Brasileiro de Sensoriamento Remoto*, Goiânia, Brazil, pp. 1629–1636, 2005.

[OLI 07] OLIVIER J., Etude spatio-temporelle de la distribution de bambous dans le Sud-ouest amazonien (Sud Pérou) – Histoire, Dynamique et Futur d'une végétation "monodominante" en forêt tropicale humide, PhD Thesis, Paul Sabatier University, Toulouse, 2007.

[SHE 16] SHEEREN D., FAUVEL M., JOSIPOVIĆ V. *et al.*, "Tree species classification in temperate forests using Formosat-2 satellite image time series", *Remote Sensing*, vol. 8, no. 9, p. 734, 2016.

8

Physiognomic Map of Natural Vegetation

8.1. Context

The method described in this chapter is based on works developed in the framework of a national project financed by the Ministry of the Environment, Energy and the Sea. In the framework of the national strategy for Biodiversity (2011–2020), the French State has set itself the objective of developing knowledge and assessment of biodiversity. Faced with the lack of precise and generalized information on the distribution and evolution of natural and semi-natural habitats in France, the Ministry in charge of ecology initiated the CarHAB program in 2011. One of the important stages of the project is to create vegetation maps using data from optical remote sensing. The objective of the realization of the maps is to achieve a geographical partitioning of the natural habitats according to a physiognomic approach of the vegetation, that is to say revealing the structure, height or biomass. The physiognomic map of natural vegetation provides a clear and homogeneous spatial framework to the field operators, even before deploying field teams.

8.2. Method

The work is based on the fusion of a mono-date image with very high spatial resolution (VHSR, a metric resolution) and an annual time series of high spatial resolution images (HSR, a decametric resolution) for mapping physiognomic natural vegetation. The first step consists of segmenting the VHSR image (1) into objects that will be characterized according to an object-oriented and hierarchical approach. The next step is to extract natural habitats based on the annual time series of the HSR images (2).

Chapter written by Samuel ALLEAUME and Sylvio LAVENTURE.

This approach is based on the fact that natural environments have a year-round crop production (apart from snow cover periods), unlike other land uses (crops, buildings, etc.). Indeed, crops are found at least once in bare soil in the year after harvest and in plowing, while urban areas have virtually zero crop production throughout the year. Subsequently, the VHSR images allow the vegetation of natural environments to be categorized in three levels of densities (3): herbaceous (open), woody mixed (semi-open) and dense woody. Then, the time series is again used to characterize herbaceous areas (lawns or grasslands) in the level of plant biomass productivity (4). Finally, all these treatments lead to the vegetation map (5). To accomplish this objective, the different parts of this chapter can be summarized as follows (Figure 8.1):

1) Segmentation or automatic division into homogeneous zones from the single-date VHSR image.

2) Extraction of natural media from the temporal series of HSR images.

3) Extraction of the degree of opening of the media, that is to say the density of natural vegetation.

4) Extraction of the productivity level of plant biomass.

5) Final vegetation map.

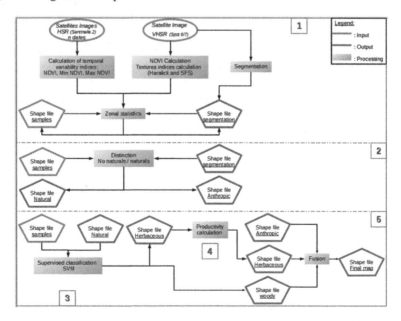

Figure 8.1. *Chain of treatments for the mapping of natural vegetation physiognomies. For a color version of the figure, see www.iste.co.uk/baghdadi/qgis2.zip*

8.2.1. *Segmentation of the VHSR mono-date image*

A segmentation consists of a grouping of pixels with the same pre-defined properties. The pixels are then grouped into regions (also called objects or segments). It creates a conceptual link with a human perception by trying to imitate the photo-interpreter's ability to delineate homogeneous zones. There are different algorithms that generate segmentation. The segments thus created, also called objects, are the basis of object-oriented classification. The segment layer thus generated will serve as a spatial reference for the following treatments and classifications.

The algorithm retained, called Mean-Shift, is based on a joint analysis of the spatial and spectral characteristics of the pixels of the image. Each pixel is characterized by its radiometric information as well as by its position. Two main parameters regulate the function of this technique: the radius range, which defines a proximity in terms of pixel value, and the spatial radius, which defines a spatial radius in number of pixels. The algorithm thus makes it possible to find a compromise between the radiometric precision and the spatial compactness (scale of the segmentation). Mean-Shift segmentation is one of the segmentation techniques currently implemented in the Orfeo Toolbox (OTB).

OTB functionality for segmentation:
- OTB>Segmentation>Segmentation (meanshift)...

8.2.2. *Calculation of temporal variability indices*

The exploitation of an annual time series of optical images makes it possible to take into account the temporal variability of the vegetations. Natural vegetation is much more spectrally stable over time than crop vegetation.

8.2.2.1. *Normalized Difference Vegetation Index (NDVI)*

The NDVI makes it possible to observe and analyze the vegetation cover from optical remote sensing images. This index reflects the very characteristic spectral signature of active vegetation. In the visible spectral band, and particularly in the red (0.6–0.7 µm), the vegetation strongly absorbs solar radiation enabling chlorophyll activity, but this effect ceases in the near-infrared (0.7–0.9 µm) where, on the contrary, the vegetation reflects strongly. Since bare soils have a small reflectance gap between red and near-infrared, the distinction between bare and covered soil is generally not a problem:

$$NDVI = \frac{\rho_{NIR} - \rho_{Red}}{\rho_{NIR} + \rho_{Red}}$$ [8.1]

Where ρ_{NIR} is the near-infrared band reflectance and ρ_{Red} is the near-red band reflectance.

The denominator is a normalization factor that partially compensates for the difference in reflectance of the surface related to the height of the sun or the angles of the images taken by the satellites. By construction, the NDVI is between −1 and 1.

OTB functionnality for NDVI calculation:
- OTB>*Feature Extraction> Radiometric Indices...*

8.2.2.2. Calculation of temporal variability indices

When time series images are corrected for atmospheric effects, it is possible to calculate the temporal variability of the NDVI in this series. This is to calculate the minimum, maximum or annual standard deviation (SD) of the time series (Figure 8.2 and 8.3).

Image date 1			Image date 2			Image date 3		
0.47	0.25	-0.03	0.46	0.77	0.90	-0.13	0.72	-0.09
0.10	0.51	-0.29	0.87	0.10	-0.52	0.32	-0.47	0.01
0.29	0.80	-0.12	0.32	0.25	-0.29	0.79	0.74	-0.42
0.74	0.30	-0.47	-0.13	0.25	-0.47	-0.19	-0.48	0.19
0.73	0.15	0.37	0.83	0.04	0.18	0.69	0.19	-0.49

MINIMUM function

Resulting image

-0.13	0.25	-0.09
0.10	-0.47	-0.52
0.29	0.25	-0.42
-0.19	-0.48	-0.47
0.69	0.04	-0.49

Figure 8.2. *Example of calculation of the minimum function on a time series of NDVI*

OTB functionnality for the minimum NDVI calculation:
- *OTB>Miscellaneous > Band Math*

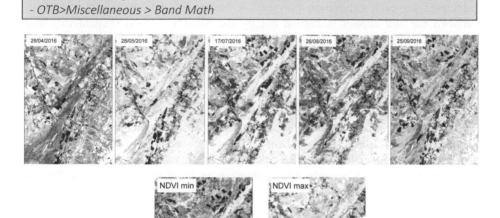

Figure 8.3. *Creation of the time indices NDVImin and NDVImax from a time series Sentinel-2*

8.2.3. *Extraction of natural vegetation using time series*

8.2.3.1. *Procedure*

A time series throughout the vegetative period (spring and summer) is used to discriminate between natural vegetation. Indeed, the minimum value of the "annual" NDVI (NDVImin) makes it possible to detect the lowest values of chlorophyll activities considered as strongly anthropized such as bare soils, urban areas or quarries. Crops areas are special because during the year they have a vegetated phase and a bare soil phase linked to harvesting and/or plowing. At the same time, natural environments maintain chlorophyll activity throughout the year, except during the winter period.

Therefore, calibration sites are used that represent the two classes to be distinguished, "crops" and "natural vegetation", to find the threshold between ever-vegetation and anthropized areas. The calibration polygons are crossed with the

NDVImin image through a zonal statistic. The average of the NDVImin is thus calculated for each calibration polygon. A search for the separability threshold between the two classes is carried out by the SEaTH method (Separability and Thresholds) detailed below. Subsequently, this threshold value is applied to attribute either the "natural vegetation" class or the "anthropized areas" class to the polygons of the segmentation.

8.2.3.2. *Classes separability using the SEaTH method*

This method makes it possible to calculate the degree of separability for a pair of classes (C1 and C2) and to provide the value of the threshold of the variable used. This method is applicable for object-oriented image classification. Two classes are thus compared across sample sites (objects) according to the "Bhattacharyya distance" B:

$$B = \frac{1}{8}(m_1 - m_2)^2 \frac{2}{\sigma_1^2 + \sigma_2^2} + \frac{1}{2}\ln\left[\frac{\sigma_1^2 + \sigma_2^2}{2\sigma_1\sigma_2}\right] \qquad [8.2]$$

For better comparability, the B value is transformed into the distance of Jeffries–Matusita or J:

$$J = 2(1 - e^{-B}) \qquad [8.3]$$

where m and σ are, respectively, the mean and the SD of the distribution of the variable for the two classes C1 and C2. The value of J thus varies between 0 and 2. The closer this value of J is to 2, the more optimal is the separability. Figure 8.4 shows three examples of separability, for probability distributions of variables for two classes.

The value of the optimal threshold is estimated for the highest value of J. It is calculated by a Bayesian statistical approach solving the equation that is in the form of a mixed Gaussian probability model for the value of x.

8.2.4. *Vegetation densities*

The classification of vegetation densities, that is to say the level of opening of the media (herbaceous, open woody, dense woody), is based on texture data from the VHSR image.

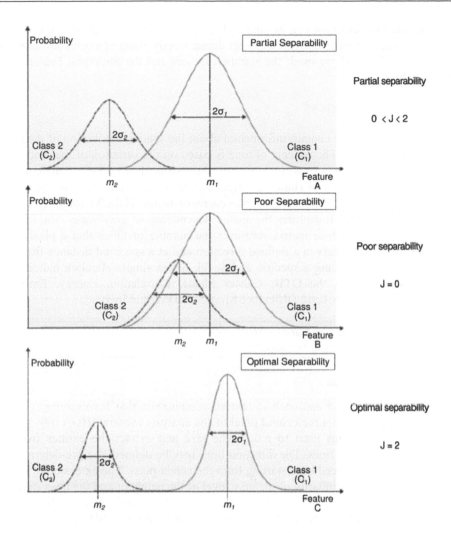

Figure 8.4. *Distribution of probabilities and values of the separability coefficient of Jeffries–Matusita (J), for an example comprising a unique attribute. For a color version of the figure, see www.iste.co.uk/baghdadi/qgis2.zip*

8.2.4.1. *Texture indices*

The very high spatial resolution allows discriminating textures for plant mosaics to be recognized. The texture refers to the variation in tone of an image and can be a good approximation of the structure of the vegetation. Texture analysis should not only distinguish herbaceous (grassland and lawn) areas from woody strata, but also separate mixed (open) woody strata from dense woody strata. Two texture indices available in the OTB are used: the Haralick indices and the Structural Feature Set (SFS) indices.

8.2.4.1.1. Haralick indices

The texture indices contain information about the spatial distribution of the tone variations for a band. The concept of tone is based on the variation of the shades of gray, whereas that of the texture concerns the spatial distribution of these tones [HAR 79]. A first group of Haralick texture indices is extracted from the matrix of levels of gray. Gray called this the co-occurrence matrix (GLCM for Gray Level Cooccurence Matrix). It captures the spatial dependency of gray-scale values in an image. The co-occurrence matrix measures the number of times that a pixel of a certain gray level appears in a defined direction and at a specified distance from its neighboring pixels having a specific gray level. Seven simple Haralick indices are thus calculated using the OTB: Cluster Shade, Correlation, Energy, Entropy, Haralick Correlation, Inverse Difference Moment (IDM) and Inertia.

OTB functionnality for the Haralick Texture Extraction:
- *OTB>Feature Extraction>Haralick Texture Extraction*

8.2.4.1.2. SFS indices

SFS is a directional approach to texture computation that involves applying a directional analysis from the central pixel of the analysis window [HUA 07]. These are statistical measures used to reduce the size and extract the entities from a histogram of direction lines. The direction lines may be defined as a pre-determined number of equally spaced lines starting from the center pixel. The extension of these lines is based on the similarity of the gray level of the neighbor and the lines starting from the central pixel in different directions. OTB calculates SFS'Length, SFS'Width, SFS'PSI, SFS'W-Mean, SFS'Ratio and SFS'SD (SD is the standard deviation).

OTB Functionnality for the SFS Texture Extraction:
- *OTB>Feature Extraction>SFS Texture Extraction*

8.2.4.2. *SVM classification*

A supervised classification is used to discriminate the three classes of vegetation densities (herbaceous, mixed woody, dense woody). The supervised classification algorithm is based on known sample sites. The technique used in this chapter is the Support Vector Machines (SVM) technique, also known as Vast Margins Separators. It consists of solving a supervised classification problem by estimating the best hyperplane in the data space as the decision boundary between two classes [OSE 16]. The hyperplane maximizes the distance between two classes of the closest points (Large margins), using support vectors (Figure 8.5).

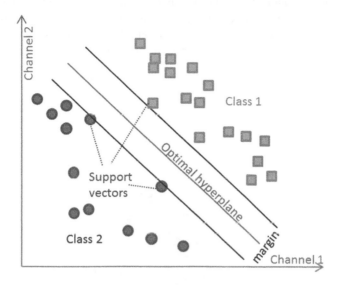

Figure 8.5. *Search for the optimal hyperplane using the SVM technique. For a color version of the figure, see www.iste.co.uk/baghdadi/qgis2.zip*

This type of supervised classification is provided by the OTB.

OTB functionality for supervised classification SVM:
- OTB>Learning>TrainImagesClassifier (svm)

8.2.5. *Maximum productivity index of herbaceous areas*

To analyze the maximum productivity of plant biomass in herbaceous areas, the segments identified as belonging to the "herbaceous" class undergo a multi-temporal analysis. Since it is necessary to distinguish the annual productivity of the herbaceous zones, the classification is based on the annual maximum NDVI (NDVImax) derived from the images of Sentinel-2.

We do not have reference samples to calibrate the NDVImax time index with field observation data. Consequently, the NDVImax is categorized in three classes with respect to the distribution histogram of the values. This approach is based on the strong assumption that maximum biomass productivity is directly correlated with annual NDVImax. The indicator is divided into three levels of productivity: low, medium and high (Figure 8.6). The threshold is calculated based on the statistical distribution of the crop production index values. In theory, in a distribution that follows a normal distribution (or bell distribution), the values of the interval between $[\mu - \sigma; \mu + \sigma]$ (where μ is the mean and σ is the standard deviation) represent 68% of the distribution values.

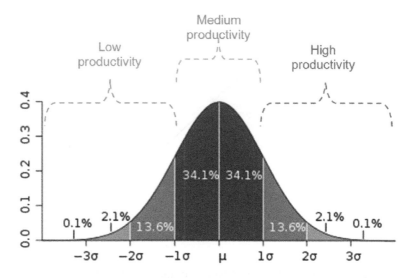

Figure 8.6. *Theoretical threshold of the three classes of vegetable production according to a normal distribution. For a color version of the figure, see www.iste.co.uk/baghdadi/qgis2.zip*

8.3. Implementation of the application

8.3.1. *Study area*

The analyzed area is located in France in the department of Isère about 10 km south-east of Grenoble between the commune of Herbeys in the north and Vaulnareys-the Bas in the south. The area of interest (latitude 45°07'N, longitude 5°48'E) is about 25 km² and covers a landscape gradient of agricultural plain in the north toward a semi-mountainous landscape to the south, reaching up to 1,000 m of altitude (Figure 8.7).

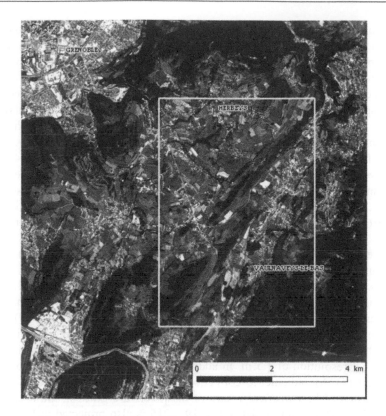

Figure 8.7. *Location of the study area. For a color version
of the figure, see www.iste.co.uk/baghdadi/qgis2.zip*

8.3.2. *Software and data*

8.3.2.1. *Software required*

The treatments in the vegetation mapping method use the basic functionalities of
the QGIS software (version 2.16) as well as the integrated image processing library
OTB, developed by CNES [CRE 17].

8.3.2.2. *Remote-sensing data*

The remote sensing images for this chapter are provided by the spatial data
infrastructure (IDS) of the Theia Continental Surface Pole (http://www.theia-land.fr).

The time series used comes from a series of Sentinel-2A images from 2016,
available for free on the site www.theia.cnes.fr. ESA's (European Space Agency)
Sentinel-2 mission, launched under the Copernicus program of the European Union,

consists of two satellites, Sentinel-2A and Sentinel-2B: the first was launched in June 2015 and the second in March 2017. Each of these satellites enables observation of the entire earth every 10 days on 13 spectral bands from the visible to the infrared means. By combining the data of the two satellites, each pixel is observed every 5 days. The images used are Level 2A data, they have been corrected for atmospheric effects and are accompanied by a good mask of clouds and cloud shadows. For this chapter, we will use the four bands with a resolution of 10 m, namely the bands B2, B3, B4 and B8 corresponding, respectively, to blue (channel 1), green (channel 2), red (channel 3) and near-infrared (channel 4). These images were reprojected in the Lambert 93 coordinate system. The five images constituting the time series date from March 26, May 28, July 17, August 26 and October 22, 2016. These images are, respectively, named S2A_20160428.tif; S2A_20160528.tif; S2A_20160717.tif; S2A_20160826.tif; S2A_20160925.tif.

The very high spatial resolution image (VHSR) used is extracted from a SPOT-7 image sampled from the entire France coverage available via Equipex Geosud (http://ids.equipex-geosud.fr). The image has a spatial resolution of 1.5 m in panchromatic mode and 6 m in multispectral mode composed of blue, green, red and near-infrared bands. This image (ORT_SPOT7_20160719_101448900_000) was cut out on the area of interest and then pre-treated by the pansharpening technique, that is the multispectral bands were fused with the panchromatic image to obtain a color image with a resolution of 1.5 m. In addition, the image was subjected to Top-of-Atmosphere (TOA) radiometric correction using the "Geosud reflectance TOA" QGIS extension. The numerical values of the pixels are given in milli-reflectance (16 bits). The SPOT-7 image dated July 19, 2016 is also projected in Lambert 93. For the chapter, this image is named SPOT7_20160719.tif. The bands are distributed as follows: red (channel 1), green (channel 2), blue (channel 3) and near-infrared (channel 4).

8.3.2.3. *Vector data*

Four vector files are provided that correspond to sample sites for calibration and validation of threshold or supervised classifications.

1) *sample_anthropic_natural.shp*: training samples required for SVM classification for the "Anthropized" and "Natural Vegetation" classes. The "Anthropized" class corresponds to urbanized areas and cultural areas, the "Natural Vegetation" class includes permanent grasslands, moors and forests.

2) *validation_anthropic_natural.shp*: validation samples to assess the distinction between the "anthropized" class and the "natural vegetation" class.

These samples are taken from a photo-interpretation. The latter used the SPOT-7 image as a support and is based on a vector layer derived from the Graphic Plot

Register (RPG). The RPG represents the vector support for declaration of land use under the Common Agricultural Policy (CAP). The RPG is data provided by the non-disseminable Agency of Service and Payment (ASP). The declaration is made annually, at the scale of one culture island on 20 classes. The attribute table of these files consists of three fields: ID corresponding to the unique identifiers of the polygons; SAMPLE text field with "A" for the class "Anthropized" and "N" for the class "Natural vegetation", which correspond, respectively, to the values 1 and 2 of the numerical field CLASS.

3) *sample_density_veg.shp*: training samples for densities classes of semi-natural vegetation.

4) *validation_density_veg.shp*: validation samples for these same density classes.

These samples come from a photo-interpretation based on the SPOT-7 image and polygons from a segmentation. The attribute table consists of three fields: ID, SAMPLE and CLASS; ID corresponds to the unique identifiers of the polygons. The text field ECHANT contains "H" for the class "Herbaceous", "M" for the class "Woody Mixed" and "D" for the class "Woody Dense" corresponding, respectively, to numerical values 3, 4 and 5 of the numerical field CLASS.

8.3.3. *Step 1: VHSR image processing*

The SPOT-7 optical image of 2016 (SPOT7_20160719.tif) is segmented into homogeneous zones.

8.3.3.1. *Image segmentation*

Procedure	Manipulation under QGIS
1. Segmentation of the VHSR image according to the Mean-Shift algorithm	In **QGIS**: • Open the image called ***SPOT7_20160719.tif*** Launch the segmentation tool in the Processing Toolbox • **Orfeo Toolbox: Segmentation > Segmentation (meanshift)** Choosing segmentation parameters: ➢ *Spatial radius*: spatial search radius in number of pixels. ➢ *Range radius*: radiometric distance in pixel value. Modify these two parameters in order to understand the effect on the segmentation. Increasing these parameters produces larger objects. ➢ *Mode convergence threshold*: the algorithm looks for modes of spatial and radiometric distribution of the pixels. A higher value of the threshold produces a reduction of the calculation times but is accompanied by a sub-segmentation.

> ➢ *Maximum number of iterations*: stops the algorithm after this number of iterations.

Post-processing options:

> ➢ *Minimum region size*: fusion of small objects (not exceeding this size) with their neighbors radiometrically closer.
> ➢ *8-neighbor connectivity*: taking account of eight connected neighbors rather than four in the default option.
> ➢ *Tile size*: the segmentation is done by block (or tile), expressed in number of pixels.

Naming the output vector file *segments.sqlite*.

Note: In its current version, the vector output must be in sqlite format, although the algorithm offers a shapefile output (.shp).

The vectorial segmentation thus produced is similar to Figure 8.8.

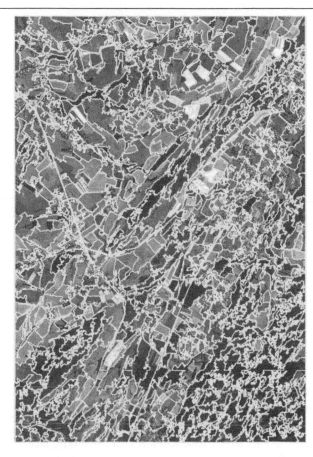

Figure 8.8. *Segmentation of the VHSR image*

Table 8.1. *Segmentation process*

8.3.3.2. *Calculation of the vegetation index NDVI on VHSR*

Procedure	Manipulation under QGIS
1. Calculation of the NDVI of the VHSR image	In the **Processing toolbox** • **Orfeo Toobox: Feature Extraction > Radiometric Indices** Assign the channel number to each spectral band (1: Red, 2: Green, 3: Blue and 4: Near Infrared) and select the radiometric number *ndvi* from the drop-down list.

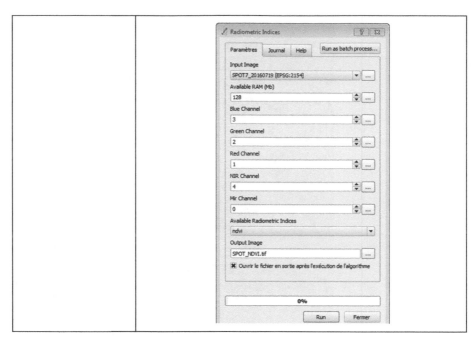

Table 8.2. *NDVI calculation*

8.3.3.3. *Calculation of texture indices*

Procedure	Manipulation under QGIS
1. Calculation of Haralick textures on the VHSR image	The simple Haralick texture indices are calculated on the red band (channel 1) of the SPOT-7 image. In the **Processing Toolbox** • **Orfeo Toolbox: Feature Extraction > Haralick Texture Extraction** Apply the basic parameters shown in the following figure. ➢ *X and Y Radius*: size of the analysis window (e.g. if X / Y radius = 2 then the size of the window will be 5x5 pixels). ➢ *X and Y Offset*: offset for computing the co-occurrence matrix (analysis direction). The output image is called ***SPOT_haralick.tif***.

	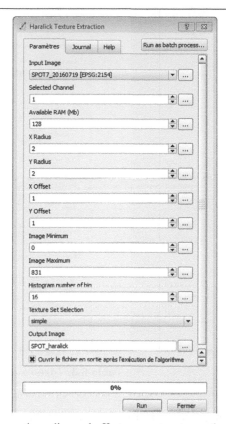 You can vary the radius and offset parameters to see the effects.
2. Calculation of SFS textures on the VHSR image	In the **Processing Toolbox** • **Orfeo toolbar: Feature Extraction > SFS Texture Extraction** Apply the basic parameters shown in the following figure. ➢ **Spectral threshold**: the maximum allowed difference between the value of the central pixel and that of a pixel of the direction line. ➢ **Spatial threshold**: the maximum length of the steering line. ➢ **Alpha and Ratio Maximum Consideration Number**: constants to adjust the value of the weighted average. The output image is called SPOT_SFS.tif.

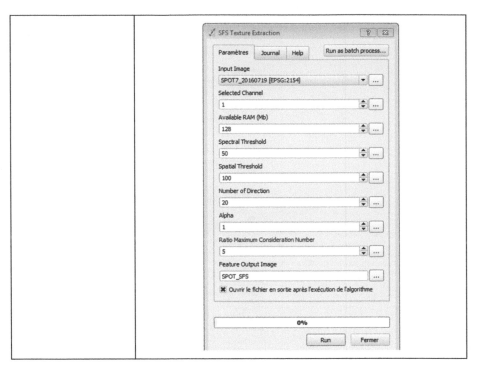

Table 8.3. *Textural index*

8.3.4. *Step 2: calculation of the variability indices on the time series*

8.3.4.1. *Calculation of the vegetation indices NDVI over time series*

Procedure	Manipulation under QGIS
1. Calculation of the NDVI on all the images of the time series Sentinel-2	Open the five images Sentinel-2: *S2A_20160428.tif; S2A_20160528.tif; S2A_20160717.tif; S2A_20160826.tif; S2A_20160925.tif* In the **Processing Toolbox** • **Orfeo Toolbox: Feature Extraction > Radiometric Indices** Assign the channel number to each spectral band (Blue, Green, Red and Near-Infrared) and select the radiometric number *ndvi* from the drop-down list.

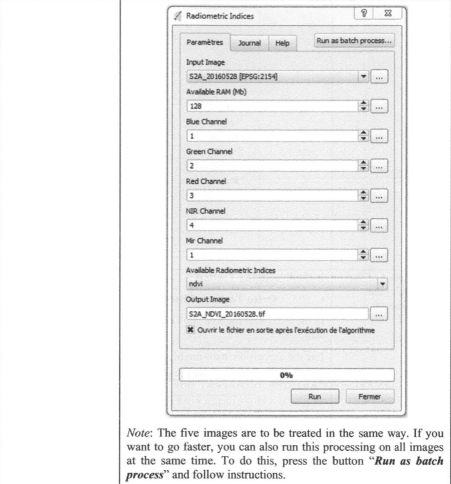

Note: The five images are to be treated in the same way. If you want to go faster, you can also run this processing on all images at the same time. To do this, press the button "***Run as batch process***" and follow instructions.

Table 8.4. *NDVI time series*

8.3.4.2. *Calculation of minimum and maximum of the temporal series*

Procedure	Manipulation under QGIS
1. Calculation of the minimum of the NDVI series of Sentinel-2	In the **Processing Toolbox** • **Orfeo Toolbox: Miscellaneous > Band Math** In the list, select the five NDVI images generated in the previous step. Use the expression: min(im1b1,im2b1,im3b1,im4b1,im5b1). Save Image under ***min_S2A_NDVI.tif***.

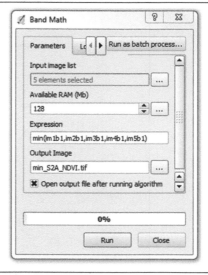

2. Calculation of the maximum of the NDVI series of Sentinel-2	In the **Processing Toolbox** • **Orfeo Toolbox: Miscellaneous > Band Math** In the list, select the five NDVI images generated in the previous step. Use the expression: max(im1b1,im2b1,im3b1,im4b1,im5b1). Save Image under ***max_S2A_NDVI.tif***. 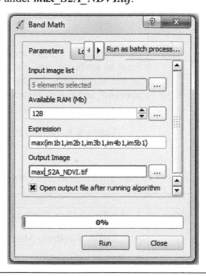

Table 8.5. *Minimum and maximum of the NDVI time series*

8.3.5. *Step 3: extraction of the natural vegetations from the time series of Sentinel-2 image by thresholding method*

Procedure	Manipulation under QGIS
1. Zonal statistic on the NDVI minimum from the time series Sentinel 2	At this level, it is necessary to extract the mean value of the minimum annual NDVI (image *min_S2A_NDVI.tif*), on the one hand for each polygon of the file that contains the training samples (*sample_anthropic_natural.shp*), and on the other hand for each object of the segmentation obtained on the VHSR image (*segments.sqlite*).

With the QGIS plugins, in the **menu**:

- *Plugins > Manage and Install Plugins*

Search "Statistics", select "Zonal Statistics plugin" and install.

In the **menu**:

- **Raster > Zonal Statistics > Zonal Statistics** ${}^{\blacksquare}\!\Sigma$

Choose file *min_S2A_NDVI.tif* as the raster layer and the sample file *asmple_anthropic_natural.shp* as the polygon layer containing the areas. Enter a significant column prefix as *min* and check the desired statistic, that is, *mean*.

2. Launch the SEaTH method using a script	Some methods used in this chapter do not exist in QGIS. However, they can be created by writing a formalized Python script that will be integrated as a toolbox.
	A script already created in Python that represents the SEaTH method is provided as a file **seath_qgis.py**.
	To add a script to QGIS:
	In the **Processing Toolbox**
	Scripts > Tools > Add scripts from file
	Select the script *seath_qgis.py*
	To run the script:
	In the **Processing Toolbox**
	Scripts > User scripts > seath qgis
	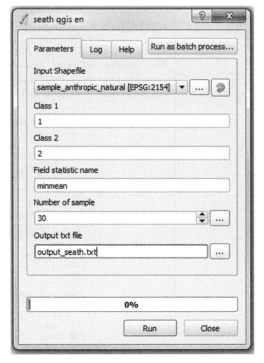
	This script creates an output file (*output_seath.txt*) providing the optimal threshold value to discriminate the two input classes: class 1, anthopic zones; class 2, natural vegetation zones. Open this output file with a text editor and note this ***threshold value***.
	*To extract the class 1, we need this index minmean < **0.473***

| 3. Apply threshold on the segmentation | In the **menu**:

• **Raster > Zonal statistics > Zonal statistics**

Choose the min_S2A_NDVI file as the raster layer and the *segments.sqlite* file as the vector layer that contains the polygons of the segmentation.

The algorithm created a new tabular field in *segments.sqlite* named *minmean*, which assigns the average value of the annual minimum of NDVI to each polygon of the segmentation.

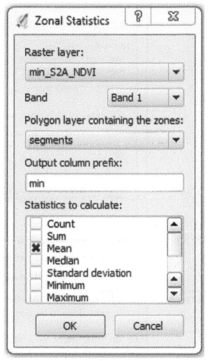

Apply the threshold value previously determined on the segments:

In the file *output_seath.txt*, it is written:

"To extract the class 1, we need this index minmean < 0.473"

Therefore, select the polygons whose *minmean* value is less than the threshold value (here 0.473) and assign them the value 1 for the "anthropic" class and the value 2 for the "natural" class in a new field called "*ant1_nat2*".

Open the attribute table (right click on the *segments.sqlite layer*, **Open Attibute Table**). |

Open Field Calculator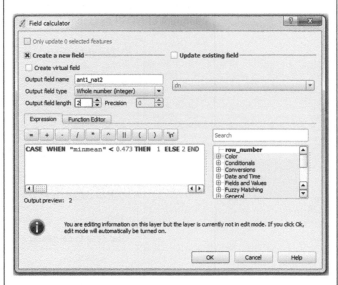

Create a new *integer* field called "*ant1_nat2*"

Use the expression

CASE WHEN "*minmean*" < 0.473 THEN 1 ELSE 2 END

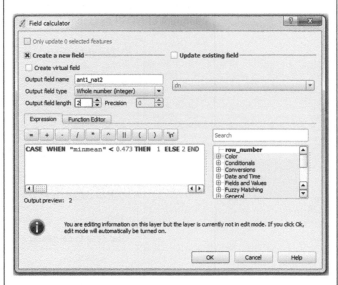

The layer has been automatically set to edit mode; therefore exit this mode after the calculation has been performed.

4. Validate the result	In order to evaluate this first result, build a matrix that allows the predicted values to be compared with validation samples.
	To do this, there is a module in the **processing toolbox** that calculates this matrix: ***ComputeConfusionMatrix.***
	This tool requests as input a raster corresponding to the map to be evaluated and a vector corresponding to the validation samples.
	Therefore, it is necessary to convert the first classification made on the *segments.sqlite* and stored in the ***ant1_nat2*** field to raster.
	The file *segments.sqlite* has to be converted beforehand and saved as a shape file: right click on the file and ***Save as … segments.shp***

In the **Processing Toolbox**:

GDAL/OGR > Conversion > Rasterize (vector to raster)

Note:

- The output file is named **anthropic_natural_map.tif.**
- Choose a resolution fine enough to reduce the bias due to rasterization (example: 2 m).

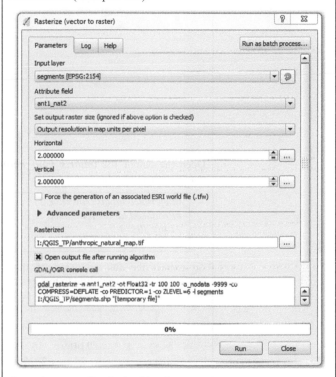

Open **ComputeConfusionMatrix**:

Orfeo Toolbox > Learning > ComputeConfusionMatrix (Vector)

- *Input Image*: raster corresponding to the map to be evaluated, ***anthropic_naturelle_map.tif***;
- *Input Reference vector data*: shapefile corresponding to the validation samples (***validation_anthropic_natural .shp***);
- *Field name*: name of the numeric field representing the classes in the evaluation file either CLASS.

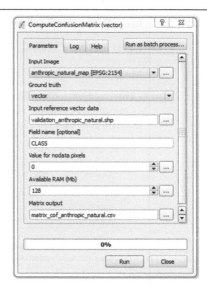

Open the output file (*matrix_conf_anthropic_natural.csv*), and read it with a text editor:

```
#Reference labels (rows) :1,2
#Produced labels (columns) :1,2
71469,30374
214,203750
```

To extract the full precision of this file, construct its own confusion matrix according to this example:

	Classes	\multicolumn{2}{c}{Classification}		Total (pxl)	Producter accuracy (%)
		1	2	Total (pxl)	Producter accuracy (%)
Reference	1	**71469**	30374	101843	0.702
	2	214	**203750**	203964	0.999
	Total (pxl)	71683	234124	305807	
	User accuracy (%)	0.997	0.870		**0.9000**

With an overall accuracy of 90%.

5. Separate layer segments into layer anthropic.shp and layer natural.shp	Create the layer *anthropic.shp*: - Open the attribute table of the layer of classified segments, *segments.sqlite*; - Launch the *selection tool* ; - Use the expression: "ant1_nat2" = 1;

- Click on 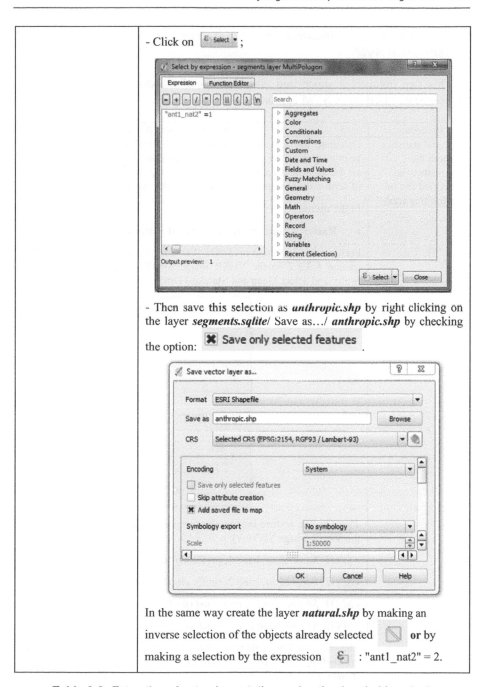;

- Then save this selection as *anthropic.shp* by right clicking on the layer *segments.sqlite*/ Save as…/ *anthropic.shp* by checking the option: **✖ Save only selected features**.

In the same way create the layer *natural.shp* by making an inverse selection of the objects already selected or by making a selection by the expression : "ant1_nat2" = 2.

Table 8.6. *Extraction of natural vegetations using the threshold method*

8.3.6. *Step 4: classification of vegetation density by supervised classification SVM*

This step involves a supervised classification of the density levels of natural environments based on an SVM approach.

Procedure	Manipulation under QGIS
1. Zonal statistics on both the segments to be classified and on the training areas	Calculate the area statistics (mean and standard deviations) on the vector layer of the natural media to be classified **natural.shp**. Then on the layer that contains the training samples for the density classes of semi-natural vegetation **sample_density_veg.shp**. In the **menu**: • **Raster > Zonal Statistics > Zonal Statistics** Perform this operation on *SPOT_NDVI.tif* (Image NDVI computed in section 8.3.3.2 on the VHSR image SPOT-7); use *ndvi* as column prefix and select Mean and Standard Deviation. Then, on the texture indices (Images calculated in section 8.3.3.3 on the VHSR image SPOT-7): - *SPOT_haralick.tif*, band 1, prefix *nrj* (for energy) - *SPOT_haralick.tif*, band 2, prefix *ent* (for entropy) - *SPOT_SFS.tif*, band 3, prefix *psi* (for PSI) - *SPOT_SFS.tif*, band 6, prefix *sd* (for Standard Deviation) The two attribute tables of the files *sample_density_veg.shp* and *natural.shp* have 10 new fields containing the predictive variables (*features*) that will be used for classification SVM: "ndvimean" "ndvistdev" "nrjmean" "nrjstdev" "entmean" "entstdev" "psimean" "psistdev" "sdmean" "sdstdev"

	Warning: - Use a prefix in lowercase. - Do not forget to do it on both vector layers, *sample_density_veg.shp* and *natural.shp*.
2. Creating a variable statistics file	- Calculate the overall statistics of the vector layer <u>to be classified</u> *natural.shp*. In the **Processing Toolbox** • **Orfeo Toolbox: Segmentation > ComputeOGRLayersFeaturesStatistics** **Warning**: - The names of the listed predictive (*features*) variables must be enclosed in quotation marks. - Manually enter the path of the .xml output file and do not click [...] because otherwise it requests an input file, which does not make sense.
3. Creating the classification model	- Classify training SVM on the vector layer of the <u>training sites</u> *sample_density_veg.shp* using the previously created xml file. In the **Processing Toolbox** • **Orfeo Toolbox: Segmentation > TrainOGRLayersClassifier**

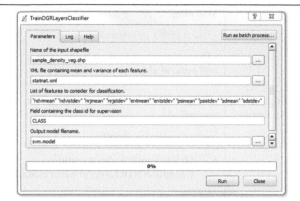

4. Assignment of predicted classes	The previous step allowed the creation of a predictive model. This model is now applied to the objects to be classified (***natural.shp***). In the **Processing Toolbox** • **Orfeo Toolbox: Segmentation > OGRLayersClassifier** 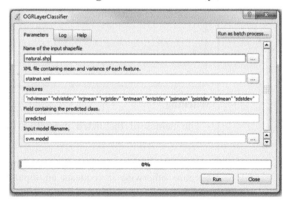 - Manually enter the path of the svm.model input file and do not click ⌐…⌐ because otherwise it requests an output file, which does not make sense. The attribute table of the layer ***natural.shp*** contains a new field (called "*predicted*") of the value of the predicted class 3, 4 or 5 corresponding, respectively, to the herbaceous, mixed woody or dense ligneous class. *Note*: - If the file ***natural.shp*** is already open in QGIS, the new field might not appear at the opening of the table. In this case, close the layer and reopen it to refresh the display.

Table 8.7. *SVM supervised classification*

8.3.7. *Step 5: extraction of the level of productivity of grasslands*

Procedure	Manipulation under QGIS
1. Zonal Statistics on the maximum of the NDVI calculated from the Sentinel-2 time series	At this level, it is a question of extracting the average values of the maximum annual NDVI for each object classified in natural environment. In the **menu**: • **Raster > Zonal Statistics > Zonal Statistics** Choose file **max_S2A_NDVI.tif** (section 8.3.4.2) as raster layer and the **natural.shp** file as polygon layer containing the natural areas. Enter a significant column prefix as **max** and check the desired statistic, that is **mean**. **Zonal Statistics** Raster layer: max_S2A_NDVI Band Band 1 Polygon layer containing the zones: natural Output column prefix: max Statistics to calculate: Count Sum ✖ Mean Median Standard deviation Minimum Maximum OK Cancel
2. Distinction of herbaceous productivity levels	The different levels of productivity are classified according to the mean and the standard deviation of the NDVImax on the set of polygons of the **natural.shp** file corresponding to the herbaceous (or natural grassland). To extract this information: - In the attribute table, select the polygons corresponding to class 3 (**herbaceous**) for the "*predicted*" field by a selection query .

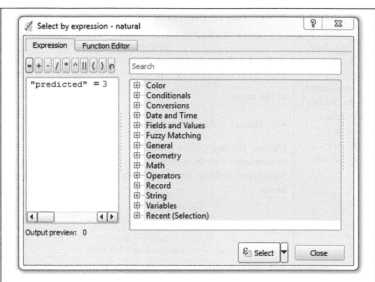

- On the main menu, select the tab **View>Statistical Summary** or click on
Σ .

A window named "Statistics Panel" will open. Select the vector layer
natural.shp, the *maxmean* field and check

✖ Selected features only

Note the mean value (μ) and the standard deviation (σ) that will be used
immediately afterward.

Calculate the intervals
-    Low productivity (LP) < μ − σ;
-    Medium productivity (MP) [μ − σ; μ + σ];
-    Very high productivity (HP) > μ + σ.

Open Field Calculator 🧮 in order to create a text field *"production"*,
which will receive the values LP, MP or HP depending on the value
maxmean. This is achieved by an expression of the following form:

CASE WHEN "predicted" = 3 and "maxmean" <0.825 THEN 'LP'

WHEN "predicted" = 3 and "maxmean" >0.835 THEN 'HP'

 WHENs "predicted" = 3 and "maxmean" >=0.825 and "maxmean"
<=0.835 THEN 'MP' END

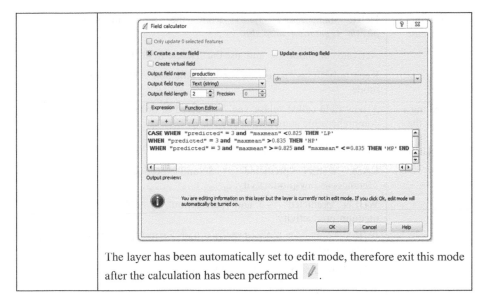

The layer has been automatically set to edit mode, therefore exit this mode after the calculation has been performed ✎ .

Table 8.8. *Productivity level of grasslands. For a color version of the table, see www.iste.co.uk/baghdadi/qgis2.zip*

8.3.8. *Step 6: final map*

Procedure	Manipulation under QGIS
1. Maps union	The final vegetation map corresponds to the set of vector layers extracted since the beginning of this chapter: ***anthropic.shp*** and ***natural.shp***. To unite these two layers and make it unique named card ***physio_vegetation_map.shp***: - Open the two layers in QGIS - In **menu: Vector > Geoprocessing Tools > Union**

2. Cleaning the attribute table	Delete unnecessary fields from the new attribute table of *physio_vegetation_map.shp*: Activate edit mode ✏ and use the "Delete field" button 🔲 . Keep only fields "*predicted*" and "*production*". Then save and exit edit mode.
3. Creating a field for the final nomenclature	A new field called "class_veg" is created in order to receive a final and comprehensible nomenclature by all. Open Field Calculator 🧮 and enter the following expression: CASE WHEN "predicted" = 3 and "production" = 'LP' THEN 'Herbaceous low-productivity' WHEN "predicted" = 3 and "production" = 'MP' THEN 'Herbaceous medium-productivity' WHEN "predicted" = 3 and "production" = 'HP' THEN 'Herbaceous high productivity' WHEN "predicted" = 4 THEN 'Woody mixed' WHEN "predicted" = 5 THEN 'Woody dense' WHEN "predicted" is NULL THEN 'Anthropic' END The layer has been automatically set to edit mode; therefore exit this mode after the calculation has been performed ✏ .
4. Symbology	- Right click on layer *physio_vegetation_map.shp* > *Properties*; - Choose tab Style; -Select ▤ Categorized ▾ ; - Choose "class_veg" as categorized column;

	- Click on the button Classify ; - Choose an appropriate symbology for each class and apply. 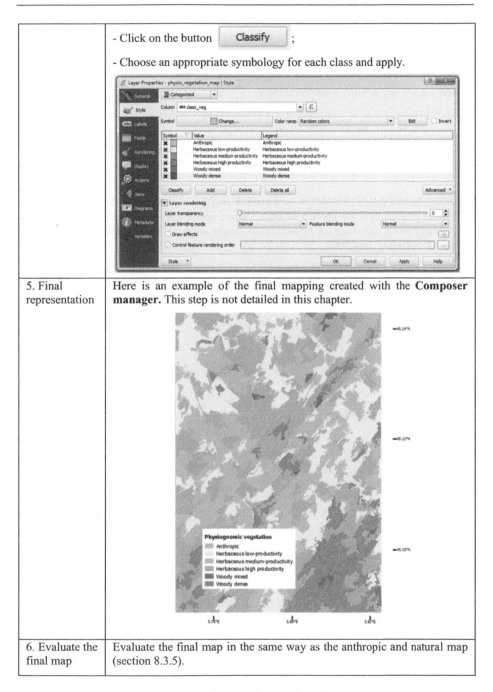
5. Final representation	Here is an example of the final mapping created with the **Composer manager.** This step is not detailed in this chapter.
6. Evaluate the final map	Evaluate the final map in the same way as the anthropic and natural map (section 8.3.5).

Table 8.9. *Final map. For a color version of the table, see www.iste.co.uk/baghdadi/qgis2.zip*

8.4. Bibliography

[CRE 17] CRESSON R., GRIZONNET M., MICHEL J., "Orfeo ToolBox Applications", in BAGHDADI N., ZRIBI M., MALLET C. (eds), *QGIS and Generic Tools*, ISTE Ltd, London and John Wiley & Sons, New York, 2017.

[HAR 79] HARALICK R.M., "Statistical and structural approaches to texture", *Proceedings of the IEEE*, vol. 67, pp. 786–804, 1979.

[HUA 07] HUANG X., ZHANG L., LI P., "Classification and extraction of spatial features in urban areas using high-resolution multispectral imagery", *IEEE Geoscience and Remote Sensing Letters*, vol. 4, no. 2, pp. 260–264, 2007.

[NUS 06] NUSSBAUM S., NIEMEYER I., CANTY M.J., "SEATH - A new tool for automated feature extraction in the context of object-based image analysis", *Proceedings of the 1st International Conference on Object-based Image Analysis (OBIA)*, ISPRS vol. XXXVI 4/C42, Salzburg, July 2006.

[OSE 16] OSE K., CORPETTI T., DEMAGISTRI L., "Multispectral Satellite Image Processing", in BAGHDADI N., ZRIBI M. (eds), *Optical remote sensing of land surface: Techniques and Methods*, ISTE Press, London and Elsevier, Oxford, 2016.

9

Object-Based Classification
for Mountainous Vegetation
Physiognomy Mapping

9.1. Definition and context

The regulatory obligation to understand, monitor and conserve their natural habitats requires European Union member states to develop long-term programs on biodiversity [LEE 03]. In response to a great lack of spatial distribution of semi-natural habitats, France has initiated a national strategy for biodiversity (2011–2020) by launching the CarHAB program (Habitats mapping) that aims to map terrestrial natural habitats with a landscape and dynamical sigmatist phytosociology approach over the whole French territory with a 1:25,000 cartographic scale. Nomenclature is based on Prodrome of French vegetation at the association level [BAR 04]. This national habitat mapping project requires a physiognomic-ecological unit's delineation from which ecologists would actually evaluate and fill habitats concretely in the field. The potential of remote sensing to automatically detect land cover on large areas represents a valuable tool to achieve the regulatory objectives of the CarHAB program.

Although environments starting at the subalpine zone represent a small size of the territory of the French country, they are extremely rich in habitats of a community interest, giving to these areas a high flora and fauna biodiversity level. Montane landscape physiognomy consists mainly of natural and semi-natural elements where ecological variability, extreme meteorological conditions and anthropogenic influence (pastoralism, forestry, etc.) make it highly heterogeneous on multi-scale levels. Thus, Very High Spatial Resolution (VHSR) optical satellite images appear indispensable

Chapter written by Vincent Thierion and Marc Lang.

for describing the physiognomic composition of mosaic-like landscape and for facilitating land cover mapping on large areas. The extreme topographic conditions of these montane areas (steep slopes and limited transportation networks) limit identification and mapping of rare and/or protected habitats.

In this chapter, we consider this mosaic-like landscape as open environments (i.e. excluding closed forests) starting at the subalpine zone. It corresponds to subalpine, alpine and nival zones up to timberline composed of mineral, herbaceous and woody gradients.

9.2. Method for detecting montane vegetation physiognomy

Automatic remote sensing detection of montane vegetation physiognomy is based on VHSR images. Despite a high spatial accuracy, these kinds of images have a high spectral heterogenity, which generally affect automatic interpretation. Their analysis dictates the use of a specific method, such as an object-oriented approach. With this method, the pixel is no longer the basic semantic unit, instead, a quite spectrally homogeneous group of pixels, so called an object.

The object-oriented method dedicated to montane open environments is based on the following nomenclature:

– Bare soil;

– Mineral grassland;

– Grassland;

– Dense moors/heaths;

– Open moors/heath;

– Dense shrubs;

– Very open coniferous forests (afforestation).

It is composed of six methodological steps (Figure 9.1):

1) Satellite image pre-processing;

2) Image segmentation;

3) Objects sampling;

4) Classification model learning;

5) Objects classification;

6) Classification accurracy computing.

The principle of each step is detailed in the following sections. This object-oriented method was adapted to open-source QGIS software in order to facilitate remote sensing and computer programming non-expert use. The majority of processing uses Orfeo Toolbox (OTB) applications that can be programmatically interfaced (C++, Python, etc.) or access via Monteverdi software [ING 09]. The step-by-step guide of each processing is described in the third section.

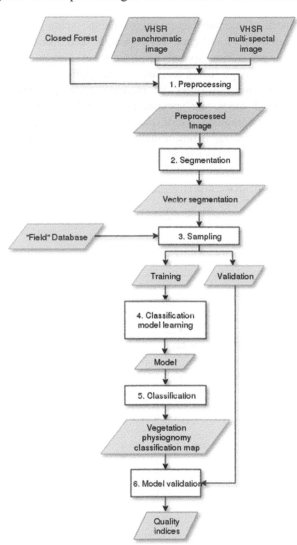

Figure 9.1. *Processing chain for montane open environments mapping. For a color version of the figure, see www.iste.co.uk/baghdadi/qgis2.zip*

9.2.1. *Satellite image pre-processing*

The pre-processing step corresponds to the preparation of the satellite image in order to apply segmentation and classification operations (Figure 9.2). In this chapter, the pre-processing stage consists of making active pixels (i.e. no masked pixels) of the image corresponding to montane open environments.

In the framework of montane open environments, spatial as well as spectral resolutions are important. Landscape heterogeneity and mineral and vegetation cover variability imply a fine spatial resolution but also a wide spectrum of wavelengths. Thus, an image fusion technique was applied at the beginning of the processing chain. Second, in order to increase spectral differences between vegetation and mineral zones, the spectral index NDVI (Normalized Difference Vegetation Index) was computed. Third, the blue band has been removed to limit correlations of visible spectral bands. Finally, after placing original image bands (except blue ones) and the NDVI band side-by-side (concatenation), we applied a mask on forested areas in order to apply following processing on interested areas, that is open environments.

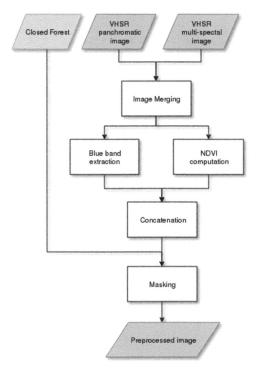

Figure 9.2. *SPOT-6 image pre-processing method. For a color version of the figure, see www.iste.co.uk/baghdadi/qgis2.zip*

9.2.1.1. *Merging panchromatic and multispectral bands*

Satellite image fusion enables us to create a new image from spectral and spatial heterogeneous images. The most common and efficient manner is a fusion of panchromatic and multispectral bands provided by the same satellite constellation, in order to combine the fine spatial resolution of a panchromatic band and spectral depth of multispectral image. Efficient image fusion involves combining panchromatic and multispectral bands with almost the same spectral depth. For instance for SPOT-6 images, used in this chapter, image fusion is a feasible solution, since the panchromatic band covers wavelengths from 0.450 to 0.745 μm (1.5 m spatial resolution) and multispectral bands cover wavelengths from 0.450 to 0.890 μm (6 m spatial resolution).

In the literature, a lot of different image fusion or pansharpening methods exist [KPA 14]. We can cite, without being exhaustive, two main categories. "Component Substitution", or RCS, is based on the substitution of one of the multispectral bands by the panchromatic one. Principal Component Analysis (PCA) can be used, for example, to transform the multispectral image and replace the first principal component with a panchromatic band. In the same vein, a transformation of visible bands of multispectral image (RGB) in intensity (I), hue (H) and saturation (S) components can be made. Another category is based on image statistics, known as Bayesian methods. In this chapter, we used a pansharpening RCS method implemented in the OTB library. A visual inspection is often useful to validate the spectral and spatial quality of the new image.

> QGIS function for panchromatic and multispectral satellite image fusion:
> - Image fusion: *Orfeo Toolbox > Geometry > Pansharpening (RCS)*[1]

9.2.1.2. *Normalized Difference Vegetation Index (NDVI)*

The NDVI is a spectral index that provides a proxy of photosynthetic activity. It enhances spectral contrast between vegetated and non-vegetated covers and provides information on biomass gradients. It corresponds to a normalized ratio between infrared and red bands of a satellite image. Over a range of wavelengths between 0.6 and 0.7 μm (red band), the radiation is highly absorbed by vegetation and reflected by mineral while between 0.7 and 0.9 μm (infrared band) radiation is highly reflected by vegetation:

$$NDVI = \frac{\rho^{NIR} - \rho^{Red}}{\rho^{NIR} + \rho^{Red}} \qquad [9.1]$$

where ρ^{NIR} is the infrared reflectance and ρ^{Red} red reflectance.

---

1 https://www.orfeo-toolbox.org/CookBook/Applications/app_Pansharpening.html.

The NDVI values are between −1 and 1. Negative values correspond to water, while values around 0 correspond to bare soil. The vegetation surfaces have NDVI values up to 0.2. Higher than 0.8, the index is quite inoperable due to saturation; however, it corresponds to dense deciduous forests with large photosynthetic activity.

> **QGIS function to compute NDVI:**
> - NDVI computing: *Orfeo Toolbox > Feature Extraction > Radiometric Indices*

9.2.1.3. *Extraction, masking and concatenation*

As explained, after blue band removing, the optical fusion image and the NDVI image have to be concatenated. Blue band corresponds to lower wavelengths of the SPOT-6 satellite image from 0.450 to 0.520 µm. This band is mainly used to detect lake and marine habitats[2].

Before concatenation, pixels that correspond to closed forests have to be masked, that is coded with an extremely high value (maximum value of image coding format or extreme value compared to non-masked pixel value). We used the French geographic database "BD Foret[®]" (version 2) provided by the IGN institute [IFN 16], which represents the most reliable database to map forested areas. Other pixels (non-masked) are preserved to execute segmentation and classification.

In order to extract the blue band and apply the "forest" mask to the fusion image, we need to separate all bands of the fusion image. Thus, we obtain one raster for each spectral band.

> **QGIS function to separate spectral bands:**
> - Spectral bands: *Orfeo Toolbox > Image Manipulation > Split Image*

Fusion image masking is made band-by-band with a simple matrix operation between rasters (image and mask). First, the thematic vector layer of the IGN database (BD Foret[®]) is rasterized.

> **QGIS function to convert vector to raster file:**
> - Vector to raster conversion: *Orfeo Toolbox > Vector Data Manipulation > Rasterization (image)*

---

2 This blue band extraction step is only for demonstration. The strong presence of water bodies in montane open environments must indeed incite the remote sensing expert to keep this spectral band for mapping larger geographical areas.

The masking operation is executed by multiplying each spectral band of fusion image (previously separated) and NDVI spectral index band with "forest" mask (0: closed forest; 1: montane open environments).

> **QGis function to apply mask:**
> - Raster math operation: *Orfeo Toolbox > Miscellaneous > Band Math*

Finally, before grouping all of the spectral bands (green, red, infrared and NDVI), pixels values have to be normalized in order to compute segmentation efficiently. Before normalization, the NDVI spectral index has pixel values between −1 and 1, while image fusion spectral bands have pixel values between 0 and 255 or 0 and 65,536, depending on image pixel format (respectively, 8 bits or 16 bits). By normalizing pixel values of each band (e.g. between 0 and 1), all pixels of all bands have the same weight during segmentation. Before the normalization process, it is necessary to calculate the mean and standard deviation of each band.

> **QGIS function to compute statistics of each spectral band:**
> - Band statistics computing: *QGIS geoalgorithms > Raster tools > Raster layer statistics*

Second, we normalize pixel values.

> **QGIS function to normalize pixel values of a spectral band:**
> - Raster math operation: *Orfeo Toolbox > Miscellaneous > Band Math*

9.2.2. *Image segmentation*

Merged images, produced in previous stages, are now spectrally and spatially well adapted to montane open environment classification objectives. Montane areas are characterized by a high mosaic-like landscape where vegetation habitats often have a size inferior to few tens of square meters. VHSR images, such as SPOT-6, have spatial resolutions adapted to capture this spatial heterogeneity. One drawback of pixel-based classification based on unique pixel analysis lies in a very heterogeneous mapping result characterized by a lot of isolated pixels, so called "salt and pepper" noise [GUO 07, YU 06]. Thus, an object-oriented approach is often better adapted to map vegetation from VHSR imagery as demonstrated many times in the literature [BAA 00, BEN 04, BOC 05, LAL 04, LAN 09, LUC 07, WEI 08].

The basic semantic primitive of classification therefore no longer corresponds to a pixel but to a set of adjacent and spectrally quite homogeneous pixels defining an object [COR 04]. Thus, the classification algorithm automatically determines membership of this object to each class of the nomenclature. The objects are obtained by the automatic segmentation of preprocessed satellite images. Numerous segmentation algorithms are described in the literature [BLA 10, HOO 96, MIK 15].

The region growing segmentation method, available in the commercial software eCognition Developer® [TRI 17], consists, for example, of iteratively merging pixels according to a spectral homogeneity criterion (scale factor) and a shape constraint. This approach belongs to the "Connected Component" segmentation algorithms family, that is based on user-defined neighbor criteria. Although the "watershed" approach analyzes the shape of the image histogram, the "Mean-Shift" segmentation method available in the OTB library with multithreading capabilities [COM 02] is particularly efficient for high dimension images (large geographic areas and high numbers of spectral bands). It is a non-parametric feature-space analysis method for partitioning multidimensional data by locating pixels that correspond to the mode or the maxima of a spectral density function. By successive iterations, this algorithm clusters neighbor pixels belonging to a same density gradient with size constraint.

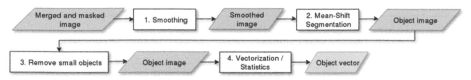

Figure 9.3. *The four steps of the Mean-Shift approach to generate objects with OrfeoToolbox library. For a color version of the figure, see www.iste.co.uk/baghdadi/qgis2.zip*

In this chapter, we used this efficient algorithm. The segmentation operation is split into four steps (Figure 9.3). It produces a vector layer of objects whose main attributes correspond to spectral statistics based on the pixel values of the merged original image (mean, standard deviation, mode, etc.):

1) mean filtering to smooth merged image;

2) *Mean-Shift* segmentation;

3) removal of small objects by neighbor analysis (user-defined threshold);

4) vectorization of objects and computation of pixel and shape statistics.

Classification accuracy is directly related to segmentation quality. In other words, object shape should be consistent with class in order to discriminate. Several published methods exist for assessing segmentation quality usually based on spatial conformity with user-defined reference objects, for example to identify under- or over-segmentation [HOO 96]. This step is not described in this chapter. In a general manner, a visual inspection of segmentation result enables us to find, step-by-step, the best parameters and best-fitting objects. In the case of large geographic areas, an efficient method consists of selecting sub-areas representative of the landscape heterogeneity of the study area. After finding the best segmentation parameters for each sub-area, the objective is then to determine the optimal parameters to segment the whole area.

QGIS function to apply mean filter:
- Smoothing filter operation: *Orfeo Toolbox > Image Filtering > Exact Large-Scale Mean-Shift segmentation, step 1 (smoothing)*

QGIS function to apply Mean-Shift segmentation:
- Segmentation operation: *Orfeo Toolbox > Segmentation > Exact Large-Scale Mean-Shift segmentation, step 2*

QGIS function to delete objects / segments under size threshold:
- Size filter operation: *Orfeo Toolbox > Segmentation > Exact Large-Scale Mean-Shift segmentation, step 3 (optional)*

QGIS function to vectorize raster segmentation:
- Vectorization operation: *Orfeo Toolbox > Segmentation > Exact Large-Scale Mean-Shift segmentation, step 4*

As explained previously, the fusion image used for segmentation has been masked. The segmentation of masked pixels, coded with an extreme value very different to no masked pixels (i.e. pixels representing montane open environments), generates fully homogeneous objects (i.e. standard deviation equal to zero). These objects can be easily removed by GIS manipulation.

9.2.3. *Sampling, learning and segmented image classification*

Once the segmentation step has been completed, objects have to be classified, that is attributed to a class of user-defined nomenclature (Figure 9.4). The choice of the nomenclature is as important as the choice of satellite images or the type of processing to apply. The definition of the nomenclature should be defined in accordance with spectral and spatial properties of the chosen satellite image. Users should find the best trade-off between images potentialities and study objectives. The following questions should be considered before automatic satellite image classification: Which is the restitution scale? Which is the expected minimal mapping unit? Which classes are needed? Which satellite images should be used?

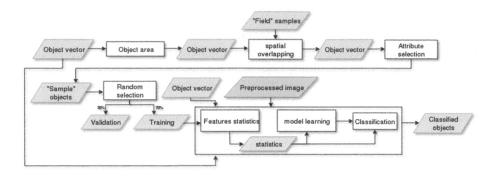

Figure 9.4. *Sampling and classification steps based on segmentation vector layer.*
For a color version of the figure, see www.iste.co.uk/baghdadi/qgis2.zip

Due to physiognomic gradients between mineral, herbaceous and woody vegetation, the mapping of montane open environments is a critical operation. Comparing the high contrasted landscape, mineral and vegetation physiognomy of montane areas resembles a succession of gradients composed of a variety of elements, from bare soils, heterogeneous grasslands and moors to pre-forested zones. A remote sensing expert would be tempted to decompose these physiognomic gradients into several heterogeneous classes. One of the biggest problems is the scale/resolution issue. For instance, the SPOT-6 image enables us to detect fine landscape elements. On the SPOT-6 satellite image, some complexes such as grassland and mineral mixture appear as homogeneous physiognomy while heterogeneous in the field. It is hence possible to define one heterogeneous class called "mineral grassland" spectrally homogeneous on SPOT-6 image and therefore possible to detect.

9.2.3.1. Sampling

Samples for classification can be retrieved from different data sources. We can use existing geographic databases such as IGN databases (e.g. BD Topo®), the *Référentiel Parcellaire Graphique* (RPG) (French Land Parcel Identification System) [CAN 14] or Corine Land Cover (CLC) [EEA 07]. For instance, CLC provides samples of large geographic areas with a large number of land cover and land use classes. In more specific cases, such as montane environments, national databases are less reliable from geographic and semantic standpoints. Thus, fieldwork and/or manual photo-interpretation are needed. One efficient solution is to perform previous processing steps (pre-processing and segmentation) and use the vector layer resulting from segmentation in order to create *in situ* or photo-interpreted samples. The segmentation vector layer can indeed be validated through this approach while samples are fully compatible with segmentation objects used for classification.

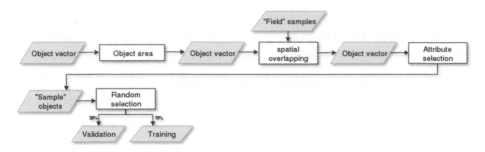

Figure 9.5. *Learning and validation samples preparation steps. For a color version of the figure, see www.iste.co.uk/baghdadi/qgis2.zip*

In this chapter, a GIS method is presented in order to use a vector layer of samples produced independently from a pre-produced segmentation, that is by manual digitization and photo-interpretation. These samples are then integrated with segmentation objects according to a user-defined spatial overlap threshold (Figure 9.5). Once integrated, the samples are split into two groups in order to avoid any misleading statistic validation, one for training and another for classification validation.

QGIS function to randomly split a sample database:
- Random sampling operation: *QGIS Geoalgorithms > Vector Selection Tools > Random selection within subsets*

9.2.3.2. *Training and classification*

The first step of the classification stage is to produce a classification model based on training samples and spectral characteristics extracted from satellite spectral bands and spectral index (e.g. NDVI), called classification primitives (Figure 9.6).

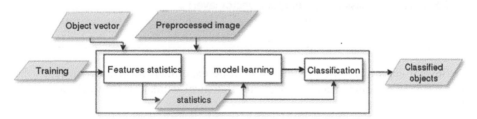

Figure 9.6. *SPOT-6 satellite image classification steps from segmentation vector layer and training samples. For a color version of the figure, see www.iste.co.uk/baghdadi/qgis2.zip*

In order to perform the classification, descriptive statistics of each primitive, that is mean, standard deviation, mode, etc., must be computed for each segmentation object and thus also training sample.

QGIS function to compute segmentation object statistics:
- Statistics computing operation: *Orfeo Toolbox > Segmentation > ComputeOGRLayersFeaturesStatistics*

The statistics dataset then leads to the classification model or classifier. The model defines a spectral signature of each land cover class, that is the mean and its standard deviation for each primitive, and defines classification rules depending on the classification method.

QGIS function to train classification model:
- Classification model train operation: *Orfeo Toolbox > Segmentation > TrainOGRLayersClassifier*

Like segmentation, there are many classification algorithms. We can cite the maximum likelihood classifier, one of the most used to study natural environments, or the Random Forest one [PAL 05], less sensitive to mislabeling training samples, as well as Support Vector Machine (SVM) particularly efficient with a low number of samples [HUA 02]. In this chapter, classification is performed with an SVM classifier based on an object-oriented approach. Unlike the pixel-based approach, only the SVM classifier is implemented in the OTB library to handle object-based classification. To experiment with other classification algorithms using an object-based approach, R (https://cran.r-project.org) or scikit (http://scikit-learn.org/), libraries can be used. Spectral statistics of segmentation-based objects can indeed be easily exported to QGIS software and used in these libraries.

QGIS function to classify segmentation vector layer:
- Classification operation: *Orfeo Toolbox > Segmentation > ComputeOGRLayersFeaturesStatistics*

Several classification algorithms should be experimented in order to reach the best result as possible. Finally, visual and statistical validation may help to choose the most efficient one.

9.2.4. *Statistical validation of classification*

The statistical validation of the vegetation physiognomic classification is the final stage of this method. Several indicators can be used to perform this validation. The first step is to compute the confusion matrix, which is a contingence matrix between classification and validation samples. From this confusion matrix, other indicators can be computed: precision, recall and F-score (or F-measure) for each class and kappa index (Figure 9.7).

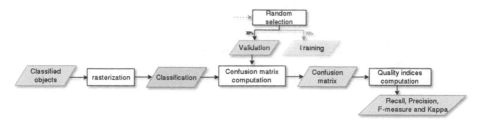

Figure 9.7. *Classification validation steps including confusion matrix and quality indicators computing (Kappa, Recall, Precision and F score). For a color version of the figure, see www.iste.co.uk/baghdadi/qgis2.zip*

In this chapter, as explained previously, a subset of samples has been kept that is not used for training, and thus the classification stage, in order to perform statistical validation.

9.2.4.1. *Confusion matrix and overall accuracy*

The confusion matrix enables us to compare the classification result with the reference dataset. Figure 9.8 corresponds to a confusion matrix. Reference/Validation samples are in rows while classification data are in columns. It is equivalent to a contingence matrix. The diagonal elements (a, b, c and d) correspond to correctly classified pixels (or areas) while all other cells are misclassifications, called confusions.

QGIS function to perform confusion matrix:
- Confusion matrix computing operation: *Orfeo Toolbox > Learning > ComputeConfusionMatrix (vector)*

Different automatic tools can produce a confusion matrix (see section 9.3.3.4) but a simple GIS layer intersection between the validation samples and classification vector layer performs the correspondences between the two datasets.

		Classification					Omission error
		Class 1	Class 2	Class 3	Class 4	Total	
Reference	Class 1	a	...	...	...	A'	1- (a /A')
	Class 2	...	b	...	...	B'	1- (b /B')
	Class 3	...	...	c	...	C'	1- (c / C')
	Class 4	...	...	...	d	D'	1- (d /D')
	Total	A	B	C	D	N	
	Commission error	1- (a /A)	1- (b /B)	1- (c / C)	1- (d /D)		

Figure 9.8. *Theoretical confusion matrix. For a color version of the figure, see www.iste.co.uk/baghdadi/qgis2.zip*

The first global quality index, measurable from the confusion matrix (Figure 9.8), is the overall accuracy:

$$\text{Overall accuracy} = \frac{(a+b+c+d)}{N} \qquad [9.2]$$

where $N = A' + B' + C' + D' = A + B + C + D$.

This first index gives an overview on global quality of the classification. It does not reflect the difference of the number of samples between classes. The kappa index responds more effectively to this frequent imbalance, by using a statistical adjustment:

$$\text{Kappa} = \frac{P_0 - P_e}{1 - P_e} \qquad [9.3]$$

where P_0 is the global classification accuracy and P_e is an estimate of the rate of agreement, called correction factor:

$$P_e = \frac{((A * A') + (B * B') + (C * C') + (D * D'))}{N^2} \qquad [9.4]$$

9.2.4.2. *Class statistical accuracy*

Several indices exist to perform class statistical accuracy. First of all, it is interesting to note that class accuracies are not the same as we observed in the confusion matrix by row or by column, that is in considering references as the classification or as the validation samples. By row, it refers to producer's accuracy, so called class precision, that is an inverse measurement of the omission errors. It corresponds to the number of correctly classified validation samples of a particular category divided by the total number of validation samples for that category:

$$\text{Producer's accuracy} = \frac{a}{A\prime}; \text{ Omission errors / Precision} = 1 - \frac{a}{A\prime}$$

By column, it refers to user's accuracy, so called recall, that is an inverse measurement of the commission errors. It corresponds to the number of correctly classified samples of a particular category divided by the total number of samples being classified as that category:

$$\text{User's accuracy} = \frac{a}{A}; \text{ Commission errors / Recall} = 1 - \frac{a}{A\prime}$$

Finally, the most representative statistical index of classification accuracy is the F-score (or F-measure), that is harmonic mean of precision and recall. It combines commission and omission errors measurement:

$$\text{F-score} = 2 * \frac{Precision*Recall}{Precision+Recall} \tag{9.5}$$

9.2.5. *Limits of the method*

The physiognomy mapping of montane open environments by remote sensing is an important challenge in the assessment of habitat conservation status and biodiversity level. Montane environments have large spatial coverage with extreme topographic conditions making their access difficult. Thus, the automatic remote sensing method can meet the needs of large scale mapping. The object-based approach is effective in preventing the obtainment of isolated pixels during the classification phase, caused by the intrinsic structure of montane landscapes. It achieves better cartographic results expected by operational services in charge of vegetation mapping (e.g. botanical conservatories).

Nevertheless, the physiognomic proximity and gradients of montane land cover classes limit the possibility of finely mapping this physiognomy. The application of this method hereafter demonstrates the efficiency of this method with a statistical accuracy of order 90%. But this validation is based on a comparison of the classification with a few numbers of samples (around 5% of segmentation objects). A visual inspection enables us to identify classification errors. For instance, we can cite the frequent case of confusion between "Grassland" and "Dense shrubs" (Figure 9.9). This confusion, frequent in vegetation classification by remote sensing, is exacerbated by the SPOT-6 acquisition date at the end of July. During this period, montane grasslands are already grazed or senescent at lower altitude while fully vegetated at higher altitude. Thus, we can observe these confusions at higher altitude. Results could certainly be improved upon by using textural or temporal primitives for classification.

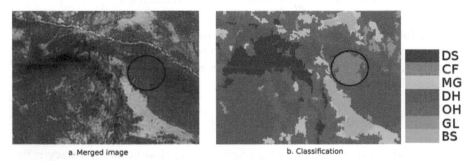

a. Merged image   b. Classification

DS
CF
MG
DH
OH
GL
BS

Figure 9.9. *Classification confusions (black circle) on fusion image (a) and classification mapping result (b). DS: dense shrubs, CF: very open coniferous forest, MG: mineral grassland, DH: dense heaths, OH: open heaths, GL: grassland and BS: bare soil. Black circle shows area that should be classified as grassland and not as shrubs. For a color version of the figure, see www.iste.co.uk/baghdadi/qgis2.zip*

In any case, a critical analysis of this mapping is essential by considering a differential, like the use of fieldwork samples. The present land cover classes are representative of the physiognomy of montane open environments. Finally, this reference map delineates homogeneous physiognomic areas with relatively accurate boundaries (spatial resolution of 1.5 m) from which ecologists can match habitats/vegetation present in the field.

9.3. Application in QGIS

This section presents an application of object-based classification for montane landscape physiognomy mapping. A SPOT-6 satellite image (one panchromatic band and four multispectral bands), acquired on July 31, 2013, was used for the different processing steps.

All image and vector processing was performed with QGIS 2.18 "Las Palmas" software. It allows to access to the image processing applications of the OTB. The statistical analysis for the validation was performed using the LibreOffice Calc 5.0.6.2 software. Finally, all processing was carried out on a Linux operating system, Ubuntu 16.04 "Xenial Xerus" distribution. Under certain configuration conditions, it is possible that there are some problems with accessing and using a few OTB applications from QGIS.

In this case, it is possible to acces directly to OTB applications (otbgui_*Tool_Name_QGIS*) from the "bin" folder of the installation folder of OTB. In a Windows operating system, double-click the application icon. In a Linux operating system, source the "otbenv.profile" file located at the root directory of the OTB installation, then execute the application.

9.3.1. *Pre-processing*

The first processing step consists of preparing the original images for classification. Several steps are necessary:

– create a fusion image from the original images;

– compute the NDVI vegetation indices;

– apply a forest mask to the images;

standardize spectral bands values.

9.3.1.1. *Image fusion*

The aim of this step is to create a multi-spectral image with a 1.5 m spatial resolution and this is done by merging two images: a single-band panchromatic image with a spatial resolution of 1.5 m and multi-spectral four-band image (blue, green, red and near-infrared) with a spatial resolution of 6 m.

Approach	QGIS steps
1. Superimposing two images	In **QGIS**, • Open the following files: o S6_panchro.TIF o S6_multi.TIF Display the **Processing Toolbox** if it is not already done. In the **menu bar**: • **Processing** > **ToolBox …** In the **Processing Toolbox**:

<table>
<tr>
<td></td>
<td>

• Click on **Orfeo Toolbox > Geometry > Superimpose sensor**
In the **Superimpose sensor** window that opens, input the parameters as displayed below:

</td>
</tr>
<tr>
<td>

2. Merging two images

</td>
<td>

At this point, the co-registration of pixels of the two images is done and merging the images can be done also.

In **QGIS**:

• Check that the following files are open:
   o   S6_panchro.TIF
   o   S6_multi_reprojected.TIF

In the **Processing Toolbox**:

• Click on **Orfeo Toolbox > Geometry > Pansharpening (rcs)**
In the **Pansharpening (rcs)** window that opens, input the parameters as displayed below:

</td>
</tr>
</table>

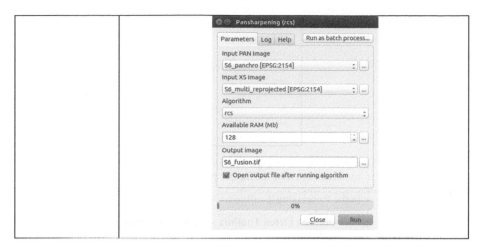

Table 9.1. *Merging of images*

9.3.1.2. *NDVI computation*

The NDVI is a normalized index whose values vary between -1 and 1. An NDVI value close to 1 expresses vegetation that is active and/or dense. Close to 0 it corresponds to bare soil and below 0 to wetlands or water bodies.

Approach	QGIS steps

1. Checking band order	In **QGIS**,
	
	• Check that the following file is open:
	    ○ `S6_fusion.TIF`
	
	In the **Layers panel**:
	
	• Double click on `S6_fusion.TIF` in order to display the Layer properties window.
	
	In the **Layer properties** window:
	
	• Click on the **Metadata** tab
	    ○ In the **Dataset Description** field we can see that:
	      ■ the first band corresponds to the blue visible range (485 nm);
	      ■ the second band corresponds to the green visible range (560 nm);
	      ■ the third band corresponds to the red visible range (660 nm);
	      ■ the fourth band corresponds to the near-infrared range (825 nm).

	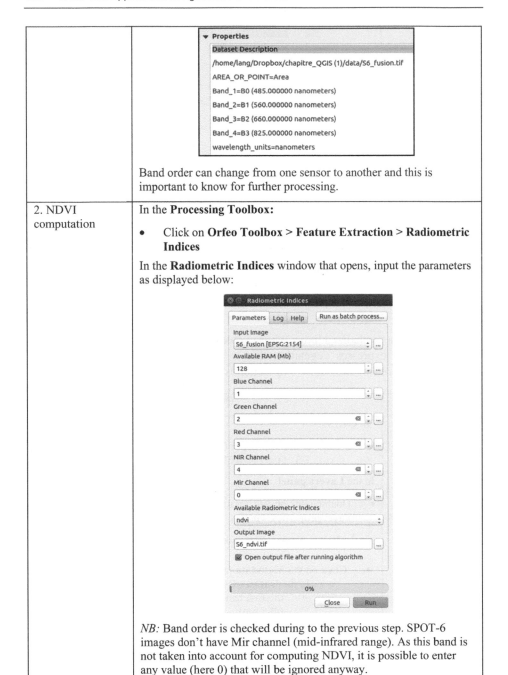
	Band order can change from one sensor to another and this is important to know for further processing.
2. NDVI computation	In the **Processing Toolbox:** • Click on **Orfeo Toolbox > Feature Extraction > Radiometric Indices** In the **Radiometric Indices** window that opens, input the parameters as displayed below: *NB:* Band order is checked during to the previous step. SPOT-6 images don't have Mir channel (mid-infrared range). As this band is not taken into account for computing NDVI, it is possible to enter any value (here 0) that will be ignored anyway.

Table 9.2. *NDVI computation*

9.3.1.3. *Applying a forest mask*

At this stage, S6_fusion.tif is available, a multi-band image of four channels and S6_ndvi.tif, a single-band image. During this step, the four bands of the fusion image are split into four separate files. Then, a forest mask is applied on the green, red, infrared and NDVI single-band images. The blue band is not useful for this application and it won't be used for the remaining step.

Approach	Steps on QGIS
1. Splitting bands	In **QGIS**, • Check that the following file is open: ○ `S6_fusion.TIF` In the **Processing Toolbox:** • Click on **Orfeo Toolbox > Image Manipulation > Split Image** In the **Split Image** window that opens, input the parameters as displayed below: • The **Output Image parameter** corresponds to the prefix used for creating output file names. • Four files are written as processing outputs: ○ `S6_fusion_split_0.tif` for blue band ○ `S6_fusion_split_1.tif` for green band ○ `S6_fusion_split_2.tif` for red band ○ `S6_fusion_split_3.tif` for infrared band
2. Forest mask rasterization	The Band Math application that will be used in the next step for forest masking on the single-band images need a raster format mask input. Hence, the forest.shp mask has to first be rasterized, that is converted to raster format (here geotiff format) from vector format (here shapefile format). In **QGIS**, • Check that the following file is open: ○ `forest.shp`

In the **Processing Toolbox:**

- Click on **Orfeo Toolbox > Vector Data Manipulation > Rasterization (image)**

In the **Rasterization (image)** window that opens, input the parameters as displayed below:

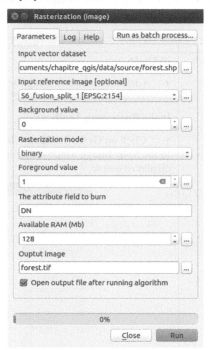

- Select any SPOT-6 image as **Input reference image** parameter, e.g. S6_fusion_split_1.tif

NB: It is crucial in the next steps that the mask and images to be masked have the same spatial resolution and extent, especially for **Band Math** application. This optional parameter allows us to respect this prerequisite by setting the output image's spatial resolution and extent from the ones belonging to a reference image.

- Select **Rasterization Mode: binary** rasterization mode

NB: In this mode, pixels inside polygons of the mask layer will be assigned the same value as indicated for the **Foreground value** parameter. The rest of the pixels will be assigned the same value as indicated for the **Background value** parameter. The alternative mode allows pixels to be assigned values corresponding to a specific field of the attribute table of the mask layer. In our case, **DN** default value of the **The attribute field to burn** parameter will be ignored.

| 3. Masking single-band images | In **QGIS**,

• Check that the following files are open:
    o   forest.tif
    o   S6_fusion_split_1.tif
    o   S6_fusion_split_2.tif
    o   S6_fusion_split_3.tif
    o   S6_ndvi.tif

NB: Blue band is not indicated here, it won't be used for the next processing steps.

• Check that `forest.tif` mask is ranked above the other single-band images:

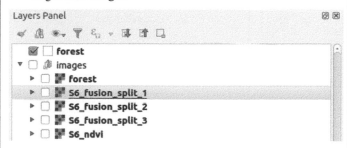

In the **Processing Toolbox:**

• Click on **Orfeo Toolbox > Miscellaneous > Band Math**

In the **Band Math** window that opens, input the parameters as displayed below:

• Select in the **Input image list** the `forest` and `S6_fusion_split_1` layers:

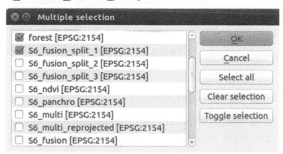

NB: Band order is important important according to the Band Math expression, that is, that is the `forest` layer must appear above the `S6_fusion_split_1` layer. If it does not, the **expression** input in the next step must be adapted (reversal of the image to be masked - im2b1 and the mask im1b1). |

- Enter the following **expression**
  `"(im1b1 == 1) ? 0 : im2b1"`. It literally means "where pixels of band 1 of image 1 (here `forest.tif`) are equal to 1, assign 0 to pixels of the output image, else assign the original values of band 1 of the image 1 (here `S6_fusion_split_1.tif`)".
- Enter `S6_fusion_split_1_masked.tif` as **Output Image:**

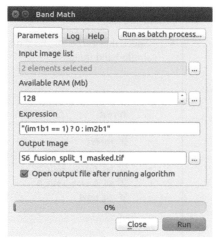

Repeat this operation for the `S6_fusion_split_2.tif`, `S6_fusion_split_3.tif` and `S6_ndvi.tif` files and respectively name the output files as:
`S6_fusion_split_2_masked.tif`
`S6_fusion_split_3_masked.tif S6_ndvi_masked.tif`

NB: Between each mask operation, the **Band Math** window needs to be closed and then reopened. Otherwise, it is possible that processing is not performed on the desired band.

Table 9.3. *Applying a forest mask*

9.3.1.4. *Bands standardization*

The band standardization operation consists of subtracting from each pixel from band B, the band mean μ and dividing it by the band standard deviation σ. It involves the following operation:

$$\frac{B-\mu}{\sigma} \qquad [9.6]$$

Firstly, statistics of each band need to be collected, and then standardization is performed band by band.

Approach	Steps on QGIS

1. Extracting band statistics	In **QGIS**:
	
	• Check that the following files are open:
	    o S6_fusion_split_1_masked.tif
	    o S6_fusion_split_2_masked.tif
	    o S6_fusion_split_3_masked.tif
	    o S6_ndvi_masked.tif
	
	In the **Processing Toolbox:**
	
	• Click on **QGIS geoalgorithms > Raster tools > Raster layer statistics**
	In the **Raster layer statistics** window that opens, input the parameters as displayed below:
	
	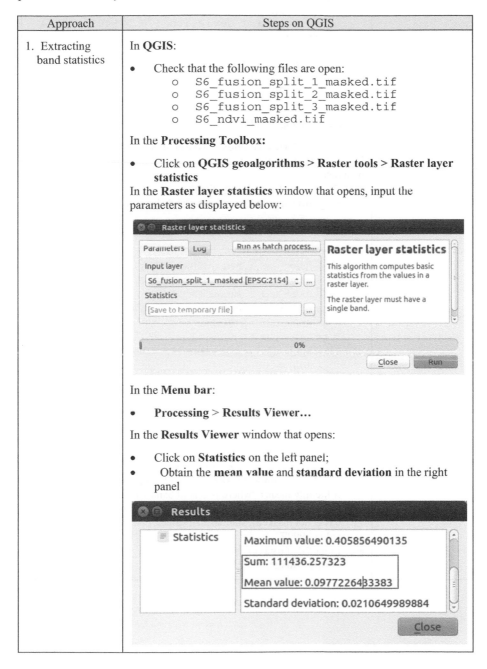
	
	In the **Menu bar**:
	
	• **Processing > Results Viewer…**
	
	In the **Results Viewer** window that opens:
	
	• Click on **Statistics** on the left panel;
	• Obtain the **mean value** and **standard deviation** in the right panel

	Repeat this operation for each band and obtain their statistics: • S6_fusion_split_2_masked.tif 　o　Mean : 0.0788979121085 　o　Standard deviation : 0.0284099745744 • S6_fusion_split_3_masked.tif 　o　Mean : 0.307375549067 　o　Standard deviation : 0.0723301106238 • S6_ndvi_masked.tif 　o　Mean : 0.575990857521 　o　Standard deviation : 0.178182842055
2. Band standardization	In the **Processing Toolbox:** • Click on **Orfeo Toolbox > Miscellaneous > Band Math** In the **Band Math** window that opens, input the following parameters: • Enter in the **Input image list** the S6_fusion_split_1_masked layer: 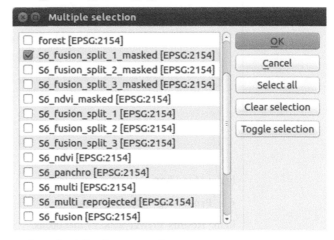 • Enter the following **expression** "(im1b1 != 0) ? (im1b1 - 0.0977226433383) / (0.0210649989884) : -9999". It literally means "where pixels of band 1 of image 1 (here S6_fusion_split_1_masked.tif) are different from 0 (i.e. not masked), standardize values, else assign a new value - 9999". • Enter S6_fusion_split_1_masked_stand.tif as **Output Image**.

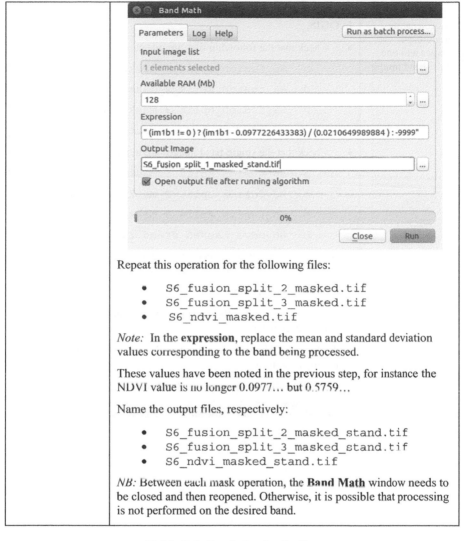

Repeat this operation for the following files:

- `S6_fusion_split_2_masked.tif`
- `S6_fusion_split_3_masked.tif`
- `S6_ndvi_masked.tif`

Note: In the **expression**, replace the mean and standard deviation values corresponding to the band being processed.

These values have been noted in the previous step, for instance the NDVI value is no longer 0.0977... but 0.5759...

Name the output files, respectively:

- `S6_fusion_split_2_masked_stand.tif`
- `S6_fusion_split_3_masked_stand.tif`
- `S6_ndvi_masked_stand.tif`

NB: Between each mask operation, the **Band Math** window needs to be closed and then reopened. Otherwise, it is possible that processing is not performed on the desired band.

Table 9.4. *Band standardization*

9.3.1.5. *Concatenation*

Now that each band has been masked and standardized, they can be stacked into one single image. Figure 9.10 shows the original bands (panchromatic and multispectral) and those generated during the pre-processing step (fusion image, NDVI and the fusion image with the forest masked).

Approach	QGIS steps
Concatenating single-band images into one multi-band image	In **QGIS**, • Check that the following files are open: o `S6_fusion_split_1_masked_stand.tif` o `S6_fusion_split_2_masked_stand.tif` o `S6_fusion_split_3_masked_stand.tif` o `S6_ndvi_masked_stand.tif` • Check that the single-band images are in the following order: o `S6_fusion_split_1_masked_stand.tif` o `S6_fusion_split_2_masked_stand.tif` o `S6_fusion_split_3_masked_stand.tif` o `S6_ndvi_masked_stand.tif` *NB:* If images are not positioned properly, they will be stacked in a different order from the one indicated above. In the **Processing Toolbox**: • Click on **Orfeo Toolbox > Image Manipulation > Images Concatenation** In the **Images Concatenation** window that opens, input the following parameters: • Input the following images in the **Input images list**: 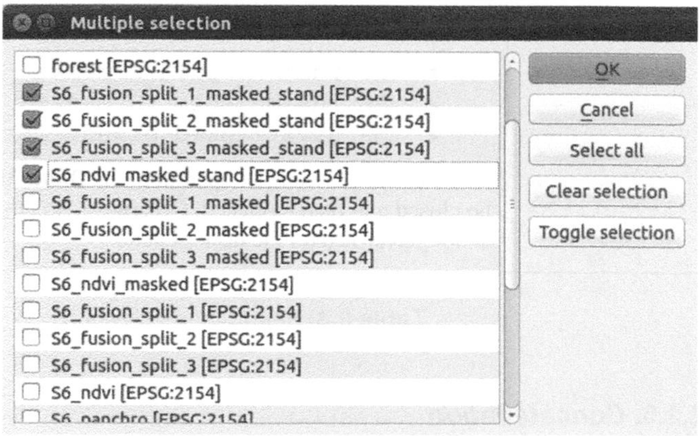 • Input `S6_concat.tif` as **Output Image**:

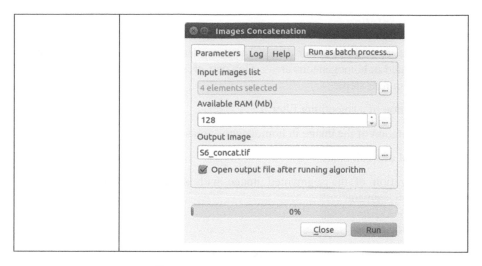

Table 9.5. *Bands concatenation*

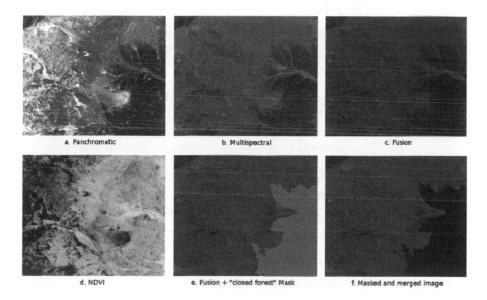

a. Panchromatic   b. Multispectral   c. Fusion

d. NDVI   e. Fusion + "closed forest" Mask   f. Masked and merged image

Figure 9.10. *Original SPOT-6 satellite images used from July 31, 2013: a) panchromatic band and b) multispectral image [RGB colored composition = Near-Infrared, Red, Green], and those generated during the preprocessing step c) image resulting from panchromatic and multispectral bands merging, d) NDVI, e) merged image with the forest mask and f) merged image without forest pixels. For a color version of the figure, see www.iste.co.uk/baghdadi/qgis2.zip*

9.3.2. *Segmentation*

At this stage, we have at our disposal one standardized multi-band image, ready to be segmented in homogeneous areas. This part presents the segmentation step that is processed in four steps:

– application of a smoothing filter;

– segmentation of the image in homogenous areas;

– removal of small areas;

– vectorization of the segmented image with statistics computation of pixel values of each segment.

9.3.2.1. *Smoothing*

During this step, the merged image is smoothed so that the detection of the homogeneous image is easier.

Approach	QGIS steps
Smoothing	In **QGIS**, • Check that the following file is open: o S6_concat.tif In the **Processing Toolbox:** • Click on **Orfeo Toolbox > Image Filtering > Exact Large-Scale Mean-Shift segmentation, step 1 (smoothing)** In the **Exact Large-Scale Mean-Shift segmentation, step 1 (smoothing)** window that opens, input the following parameters: • **Input image**: S6_concat.tif • **Spatial radius**: 15 • **Range radius**: 0.25 • **Mode convergence threshold**: 0,1 (default value) • **Maximum number of iterations**: 100 • **Range radius coefficient**: 0 • **Mode search**: disable • **Filtered output**: S6_smoothed.tif • **Spatial image**: S6_smoothed_pos.tif

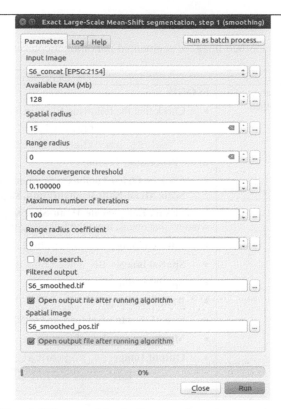

NB: When 0.25 is entered for the **Range radius** parameter, the displayed value is 0, but it is indeed 0.25 that will be taken into account for processing. It is possible to check it in the **Log** tab after running the algorithm.

Table 9.6. *Smoothing*

9.3.2.2. *Segmentation*

The smoothed image is segmented in homogeneous areas during this step.

Approach	QGIS steps

Segmentation of the smoothed image	In **QGIS**,
	
	• Check that the following files are open:
	    o S6_smoothed.tif
	    o S6_smoothed_pos.tif
	
	In the **Processing Toolbox:**
	
	• Click on **Orfeo Toolbox > Segmentation > Exact Large-Scale Mean-Shift segmentation, step 2**
	
	In the **Exact Large-Scale Mean-Shift segmentation, step 2** window that opens, input the following parameters:
	
	• **Filtered image**: S6_smoothed.tif
	• **Spatial image:** S6_smoothed_pos.tif
	• **Spatial radius**: 7
	• **Range radius**: 0.08
	• **Size of tiles in pixel** for x and y: 500
	• **Maximum number of iterations**: 100
	• **Output Image**: S6_segmentation_step2.tif

NB: When 0.08 is input for the **Range radius** parameter, the displayed value is 0, but it is indeed 0.08 that will be taken into account for processing. It is possible to check it in the **Log** tab after running the algorithm.

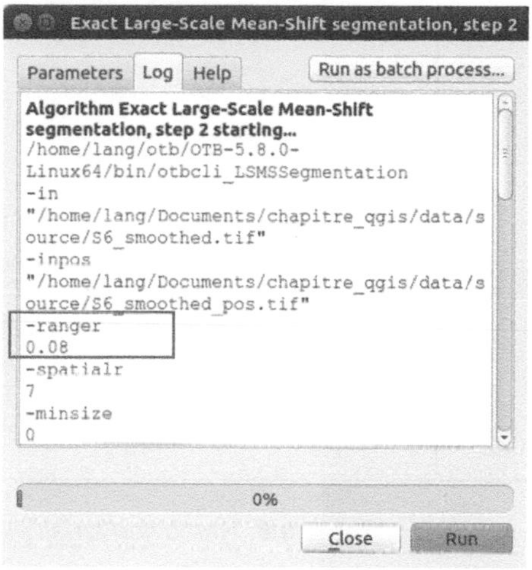

Table 9.7. *Segmentation*

9.3.2.3. *Removing small elements*

Segments that are too small are merged with the neighboring segment whose radiometry is the closest from itself.

Approach	QGIS steps

Removing small elements	In **QGIS**,
	
	• Check that the following files are open:
	
	   o S6_concat.tif
	
	   o S6_segmentation_step2.tif
	
	In the **Processing Toolbox:**
	
	• Click on **Orfeo Toolbox > Segmentation > Exact Large-Scale Mean-Shift segmentation, step 3 (optional)**
	
	In the **Exact Large-Scale Mean-Shift segmentation, step 3 (optional)** window that opens, input the parameters as displayed below:

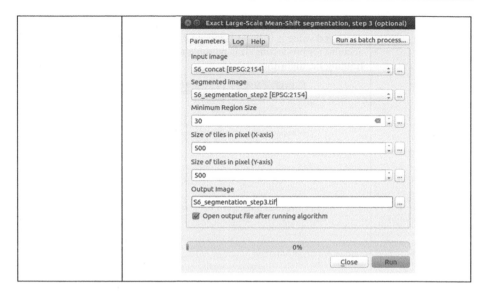

Table 9.8. *Removing small elements*

9.3.2.4. *Vectorization*

The segmented image is vectorized during this step. Mean and standard deviation of pixels within each segment are also computed for each of the four bands (Figure 9.11).

Approach	QGIS steps
1. Segmentation vectorization	In **QGIS**, • Check that the following files are open: o `S6_concat.tif` o `S6_segmentation_step3.tif` In the **Processing Toolbox:** • Click on **Orfeo Toolbox > Segmentation > Exact Large-Scale Mean-Shift segmentation, step 4** In the **Exact Large-Scale Mean-Shift segmentation, step 4** window that opens, input the parameters as displayed below:

In the **attribute table** (right click on the layer to access it), note that mean and standard deviation statistics were computed for each of the four bands (meanb0, meanb1, meanb2, meanb3, varb0, varb1, varb2, varb3).

| 2. Deleting segments corresponding to the masked areas | The resulting segmentation includes three segments corresponding to masked areas. Those segments are not meant to be classified. They instead need to be removed. |

In **QGIS**:

*   **Click** on S6_segmentation.shp in the **Layers Panel**;

*   Using the **Select Features** tool 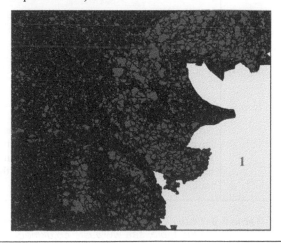, select the three segments (1, 2 and 3 in the following figure) corresponding to masked areas and that are in yellow in the figure (hold SHIFT key for multiple selection).

In the S6_segmentation.shp **Attribute table** (right click on layer to see it):

- Click on **Invert selection** .

All segments except the three selected previously are now selected.

In the **Layers panel**:

- **Right click** on the S6_segmentation.shp layer, then click on **Save As...**

In the **Save vector layer as ...** window that opens, input the following parameters:

- Enter S6_segmentation_without_forest.shp as **File Name** output;

- Select RGF93 **CRS;**

- Tick the **save only selected features** box;

- Leave the default value for the rest of options;

- Click on **OK.**

NB: In the case where the number of segments to be deleted would have been more greater, it would have been easier to select them according to their bands' mean values, which is far lower than the masked segments. This can be done using the **select by expression** tool.

Table 9.9. *Obtaining segmentation in vector format. For a color version of the table, see www.iste.co.uk/baghdadi/qgis2.zip*

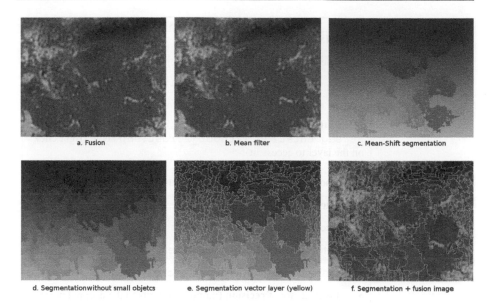

a. Fusion   b. Mean filter   c. Mean-Shift segmentation

d. Segmentationwithout small objetcs   e. Segmentation vector layer (yellow)   f. Segmentation + fusion image

Figure 9.11. *Images and vector layers used for segmentation a) merged image, RGB composite = Near-Infrared, Red and Green, and those produced during processing with b) the smoothed image, c) the segmentation produced with the Mean-Shift algorithm (here segments can be distinguished from each other with different shades of green), d) the segmentation without little segments, e) the vectorized segmentation and f) the vectorized segmentation and superimposed on a). For a color version of the figure, see www.iste.co.uk/baghdadi/qgis2.zip*

9.3.3. Classification

Now that the merged image has been segmented, segments can be classified. For this purpose, two groups of segments need to be selected: one that will be used to train a classification model and another to estimate the quality of the resulting classification map.

9.3.3.1. *Preparing samples*

We have a vector format sample database produced by both visual photo-interpretation and field trips at our disposal. First, segments corresponding to the samples need to be extracted and then separated into two sets: 30% for validation and 70% for training.

Approach	QGIS steps

1. Computing segments area	In **QGIS**,
	
	• Check that the following files are open:
	
	   o S6_segmentation_without_forest.shp
	
	   o sample.shp
	
	In the **attribute table** of the S6_segmentation_without_forest.shp layer (right click on the layer to access it):
	
	• Open the **field calculator** by clicking on .
	
	In the **field calculator**:
	
	• Check the **Create a new field** box;
	
	• Enter the **Output field name** : Area;
	
	• Define the **Output field type** : Decimal number (real);
	
	• Enter 2 for **Precision** (number of digits to the decimal point);
	
	• One the right panel, double-click on **Geometry > $area**;
	
	• The area is computed by default in the measurement projection system unit, here, in meters;
	
	• Click **OK**.

2. Union between the segmentation and the sample layers	In **QGIS**,
	
	• Check that the following files are open:
	
	    o S6_segmentation_without_forest.shp
	
	    o sample.shp
	
	In the **menu bar**:
	
	• Click on **Vector >Geoprocessing Tools> Union**
	
	In the **Union** window that opens, input the parameters as displayed below:
	
	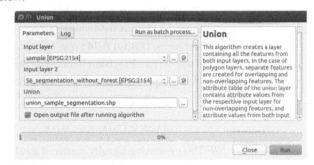

3. Computing the proportion of segments' area covered by a polygon of the sample layer	In **QGIS**,
	
	• Check that the following file is open:
	
	    o Union_sample_segmentation.shp
	
	• Open the attribute table of the following layer:
	
	    o Union_sample_segmentation.shp (right click on the layer)
	
	In the **attribute table**,
	
	• There are fields of the attribute table of both sample.shp and segmentation_without_forest.shp layer;
	
	• There are three different entity types:
	
	    o Those that have values only for the fields of the sample.shp layer. There are entities that correspond to polygons from the sample.shp layer that intersects no segment from the segmentation_without_forest.shp layer. In our case, that cannot be because all sample entities are within the image extent. Hence, these are entities that do not have geometry.
	
	    o Those that have values only for the fields of the segmentation_without_forest.shp layer. These are entities that correspond to segments from the segmentation_ without_forest.shp layer that does not intersect the polygon from the sample.shp layer.

	o Those that have values for the fields of the segmentation_without_forest.shp layer and for the fields of the sample.shp layer. They correspond to the intersection between the polygons of the two layers. It is possible to determine the area proportion recovered by the samples for each segment by calculating the ratio between the area of this type of entity and the area of the segments of origin (computed previously on the field Area). • Open the **Field Calculator** by clicking on 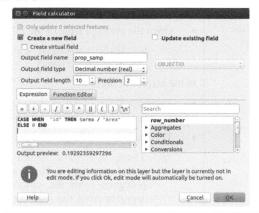. In the **Field Calculator**: • Check the **Create a new field** box; • Enter the **Output field name** : prop_samp; • Define the **Output field type**: Decimal number (real); • Enter 2 for **Precision** (number of digits to the decimal point) ; • In the **Expression** panel, enter the following formula: CASE WHEN "id" THEN $area / "Area" ELSE 0 END *NB:* This expression literally means: if a value is found in the field "id", compute the ratio between the entity area and the value of the "Area" field, else put 0. • Click on **OK**: 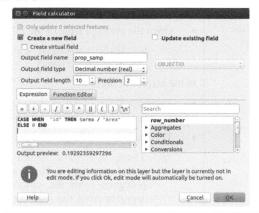
4. Assigning segments that are overlapped by more than 70% by samples, the values of the latter in the attribute table	Open the **attribute table** of the Union_sample_segmentation.shp layer (right click to access to it). First, entities that have a "**prop_samp**" ratio greater than 0.70 are selected. In the **Attribute table**:

- Click on the  icon (select features using an expression);

- Enter the following **expression:**

  "prop_samp" >= 0.7

- Click on **Select**:

Now that entities are selected, in the **Menu bar**:

- **Vector > Data Management Tools > Join attributes by location**

In the **Join attributes by location** window that opens, input the parameters as displayed below:

Now, we have access to samples whose delineations correspond to the segments of the segmentation that was carried out previously.

5. Randomly selecting samples	The samples database now needs to be split into two sets: one for training the classification model and another for validating it.
	In the **Processing Toolbox:**
	• Click on **QGIS Geoalgorithms > Vector Selection Tools > Random selection within subsets**
	In the **Random selection within subsets** window that opens, input the parameters as displayed below:
	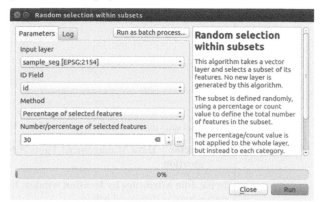
	Now that 30% of samples have been selected for each class, it is possible to save a new file of samples for validating the classification.
	In the **Layers panel:**
	• **Right click** on the `sample_seg` layer
	• Click on **Save as ...**
	In the **Save vector layer as ...** window that opens, input the following parameters:
	• Enter `sample_seg_vali.shp` as output **File name**;
	• Enter `RGF93` as **CRS**;
	• Tick the **Save only selected features** box;
	• Select the following fields to export: `label`, `nbPixels`, `meanb0`, `meanb1`, `meanb2`, `meanb3`, `varb0`, `varb1`, `varb2`, `varb3`, `Area`, `num`, `class`, `id`
	NB: This step is optional. It allows us to remove from the attribute table the unnecessary fields that were created during previous steps, in particular during the union between segmentation and sample layers.
	• Click on **OK**.

It is now necessary to create the sample training file with the 70% of samples left.

In the **attribute table** of the sample_seg layer (right click on the layer to access it):

- Click on the **Invert the selection** icon.

In the **Layers panel**:
- **Right click** on the sample_seg layer
- Click on **Save as ...**

In the **Save vector layer as ...** window that opens, input the following parameters:

- Enter sample_seg_train.shp as output **File name**;
- Enter RGF93 as **CRS**;
- Check the **Save only selected features** box;
- Select the following **fields to export**: label, nbPixels, meanb0, meanb1, meanb2, meanb3, varb0, varb1, varb2, varb3, Area, num, class, id, prop_samp

NB: This step is optional. It allows us to remove from the attribute table the unnecessary fields that were created during previously steps, in particular during the union between segmentation and sample layers.
- Click on **OK**.

Table 9.10. *Samples preparation*

9.3.3.2. *Computing statistics*

Now that the sample training and sample validation files are ready, statistics on each of the primitive need to be computed. For instance, for the "meanb0" feature (the mean value of the pixels of each segment or sample for the green band), the operation consists of computing the mean and standard deviation of all segment (or sample) "meanb0" values. These statistics files are necessary for the classification steps.

Approach	QGIS steps
1. Computing primitive statistics for training samples	In **QGIS**: • Check that the following file is open: o `sample_seg_train.shp` In the **Processing Toolbox**: • Click on **Orfeo Toolbox > Segmentation > ComputeOGRLayersFeaturesStatistics** In the **ComputeOGRLayersFeaturesStatistics** window that opens, input the parameters as displayed below: 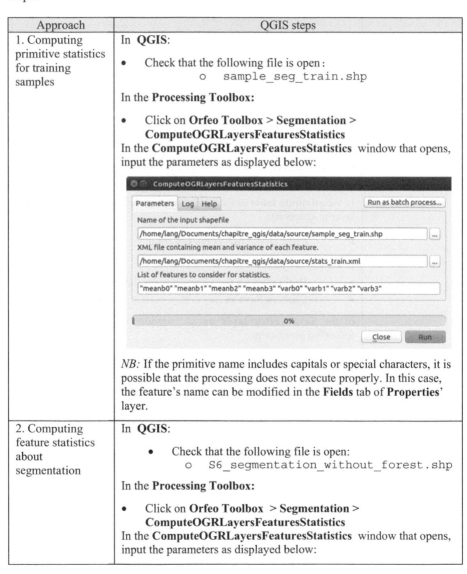 *NB:* If the primitive name includes capitals or special characters, it is possible that the processing does not execute properly. In this case, the feature's name can be modified in the **Fields** tab of **Properties'** layer.
2. Computing feature statistics about segmentation	In **QGIS**: • Check that the following file is open: o `S6_segmentation_without_forest.shp` In the **Processing Toolbox**: • Click on **Orfeo Toolbox > Segmentation > ComputeOGRLayersFeaturesStatistics** In the **ComputeOGRLayersFeaturesStatistics** window that opens, input the parameters as displayed below:

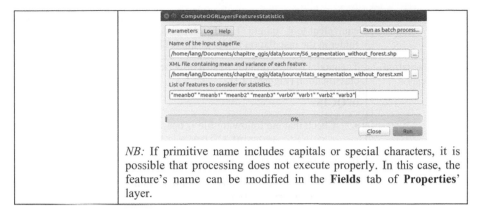

NB: If primitive name includes capitals or special characters, it is possible that processing does not execute properly. In this case, the feature's name can be modified in the **Fields** tab of **Properties**' layer.

Table 9.11. *Computing feature statistics about segmentation for classification*

9.3.3.3. *Classification*

Approach	QGIS steps
1. Model training with sample training	In **QGIS**: • Check that the following file is open o `sample_seg_train.shp` In the **Processing Toolbox:** • Click on **Orfeo Toolbox > Segmentation > TrainOGRLayersClassifier** In the **TrainOGRLayersClassifier** window that opens, input the parameters as displayed below: 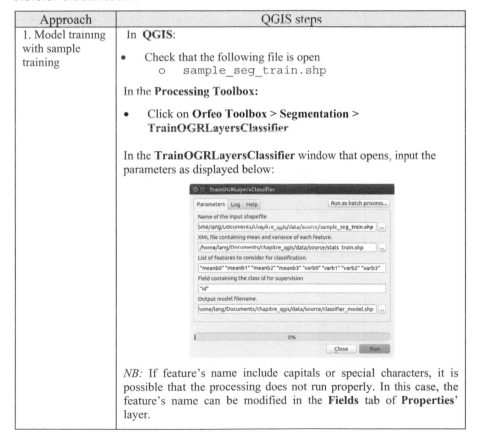 NB: If feature's name include capitals or special characters, it is possible that the processing does not run properly. In this case, the feature's name can be modified in the **Fields** tab of **Properties**' layer.

	NB: If the TrainOGRLayersClassifier does not work, use the equivalent tool "otbgui_TrainVectorClassifier" available in the "bin" folder in the installation folder of the OTB.
2. Segmentation classification	The model trained previously can now be used for segmentation classification (Figure 9.12). In **QGIS**: • Check that the following file is open: o `S6_segmentation_without_forest.shp` In the **Processing Toolbox**: • Click on **Orfeo Toolbox > Segmentation > OGRLayerClassifier** In the **OGRLayerClassifier** window that opens, input the parameters as displayed below: 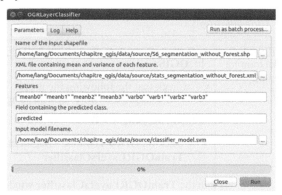 *NB:* If feature's name includes capitals or special characters, it is possible that processing does not run properly. In this case, the feature's name can be modified in the **Fields** tab of **Properties'** layer. *NB:* If the OGRLayerClassifier does not work, use the equivalent tool "otbgui_VectorClassifier" available in the "bin" folder in the installation folder of the OTB. In the **attribute table** of `S6_segmentation_without_forest.shp` layer: • Note that a new field `predicted` was created. It contains class values predicted by the classification model. *NB:* It is possible that the field appears only after having closed the file and reopened it in QGIS.

Table 9.12. *Classification*

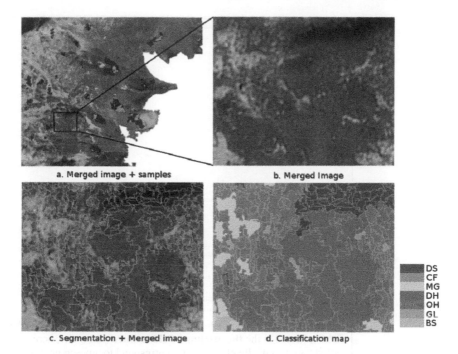

Figure 9.12. *Data used a) training samples in blue and validation samples in green, b) merged image and c) segmentation at vector format and data produced during the classification step (d) the classification map. DS: dense shrubs, CF: very open coniferous forest, MG: mineral grassland, DH: dense heaths, OH: open heaths, GL: grassland and BS: bare soil. For a color version of the figure, see www.iste.co.uk/ baghdadi/qgis2.zip*

9.3.3.4. *Validation*

Approach	QGIS steps
1. Rasterization of the segmentation that contains predicted values	In **QGIS**: • Check that the following file is open: o `sample_seg_train.shp` In the **Processing Toolbox:** • Click on **Orfeo Toolbox** > Vector Data Manipulation > Rasterization (Image) In the **Rasterization (Image)** window that opens, input the parameters as displayed below:

	NB: The **attribute** mode allows us to assign to pixels the value of the field indicated in the **the attribute field to burn** parameter, here called "predicted". Pixels that don't correspond to any segment will be assigned a value equal to the one that is indicated for the **Background value** parameter. The **foreground value** is only taken into account in **Binary** mode and thus will be ignored in this case.
2. Classification map visualization	In the **Layers panel**, double-click on the classification layer. In the **Layer properties** layer window that opens: • Select the **Style** tab; • Select "singleband pseudocolor" **render type**; • Enter 1 and 7 as the **Min** and **Max** values; • Select the "equal interval" **Mode**; • Enter 7 for the class **Number**; • Click on **Classify**. Seven classes, from 1 to 7 should appear; • Choose a color for each class by double-clicking on the color. A palette color example is displayed below:

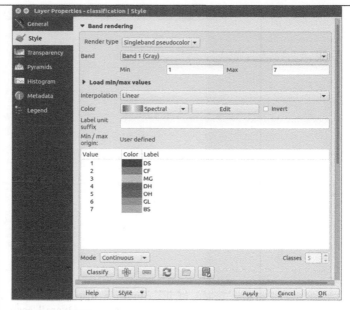

- Click on **OK**;

- You can now visualize the result in **QGIS** (Figure 9.12).

3. Computing the confusion matrix of the classification with validation samples	In **QGIS**:
	
	• Check that the following files are open:
	
	    ○ `classification.tif`
	
	    ○ `echantillons_seg_vali.shp`
	
	In the **Processing Toolbox:**
	
	• Click on **Orfeo Toolbox > Learning > ComputeConfusionMatrix (vector)**
	
	In the **ComputeConfusionMatrix (vector)** window that opens, input the parameters as displayed below:

Raw results of the confusion matrix are stored in a csv file. The accuracy indexes such as **Kappa**, **Overall accuracy** or **producer's accuracy** (or Recall) and **user's accuracy** (or Precision) are not exported to the csv file but you can check them in the log tab of the confusion matrix processing tool.

They can also be computed from the confusion matrix as explained hereafter.

4. Computing accuracy indexes Kappa, Recall, Precision and F-score	From **LibreOffice Calc**:

- Open file `mat_conf.csv`

The importation module is now open. Do not modify default options and check if the **Separator Options** is a **comma.** Rows and columns of the matrix must be clearly separated with a value for each cell.

After importation, modify the original matrix to obtain the same one as shown hereafter, by deleting rows and columns. Then, save the file (e.g. in .xlsx format):

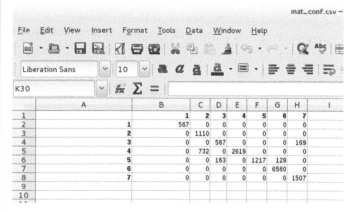

As mentioned previously, nomenclature is as following:

1. Very open coniferous forests (afforestation)

2. Dense shrubs

3. Grassland

4. Dense moors/heaths

5. Open moors/heath

6. Mineral grassland

7. Bare soil

Once the **confusion matrix is** ready, compute the **Totals**:

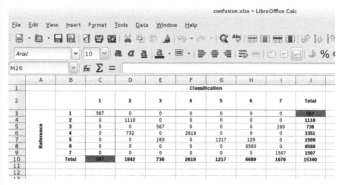

NB: From this step, it is possible to compute row or/and column accuracy rates, that is respectively based on references or classification.

It is now possible to compute **producer's** and **user's accuracy**:

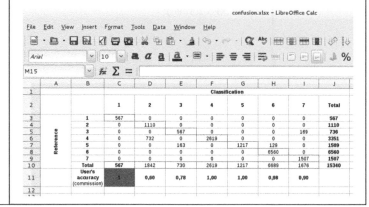

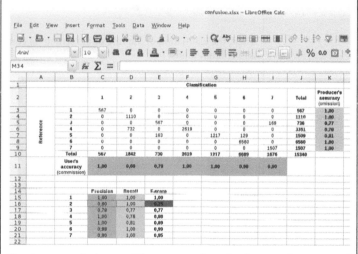

As explained in the first section of this chapter, the best accuracy index is the harmonic mean or F-score (or F-measure) considering **commission** as well as **omission errors**. In another cells zone, copy **user's accuracy** (Precision) and **producer's accuracy** (Recall) and compute F-score:

$$\text{F-score} = 2 * \frac{Precision * Recall}{Precision + Recall}$$

For instance, it is possible to note that the best performance of classification is for "Very open coniferous forests". Conversely, "Dense shrubs" has a lower accuracy.

NB: These statistical results have to be attenuated and completed with a visual inspection of the classification. Only 30% of original samples have indeed been used for this validation.

The last step is the computation of global accuracy indexes such as **Overall Accuracy** and **Kappa**.

Concerning **Overall Accuracy**, we add up the values of the diagonal divided by the total of the rows/columns totals, that is well-classified pixels divided by the whole of validation sample pixels. In this case, **Overall Accuracy** is equal to 92%.

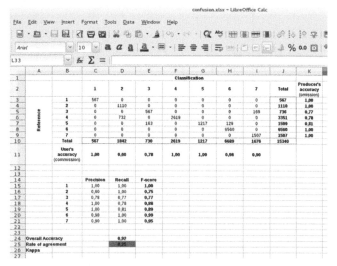

Then, the estimation of the rate of agreement is computed using row and column totals:

Finally, kappa index is computed using the following formula computation:

$$Kappa = \frac{P_0 - P_e}{1 - P_e}$$

where P_0 is the overall accuracy of the classification and P_e is an estimate of the rate of agreement, called correction factor.

$$\text{Kappa} = \frac{0.92 - 0.25}{1 - 0.25} = 0.9$$

confusion.xlsx − LibreOffice Calc

File Edit View Insert Format Tools Data Window Help

		Classification								
		1	2	3	4	5	6	7	Total	Producer's accuracy (omission)
	1	567	0	0	0	0	0	0	567	1,00
	2	0	1110	0	0	0	0	0	1110	1,00
Reference	3	0	0	567	0	0	0	109	736	0,77
	4	0	732	0	2619	0	0	0	3351	0,78
	5	0	0	163	0	1217	129	0	1509	0,81
	6	0	0	0	0	0	6560	0	6560	1,00
	7	0	0	0	0	0	0	1507	1507	1,00
	Total	567	1842	730	2619	1217	6689	1676	15340	
	User's accuracy (commission)	1,00	0,80	0,78	1,00	1,00	0,98	0,90		

		Precision	Recall	F-score
	1	1,00	1,00	1,00
	2	0,60	1,00	0,75
	3	0,78	0,77	0,77
	4	1,00	0,78	0,88
	5	1,00	0,81	0,89
	6	0,98	1,00	0,99
	7	0,90	1,00	0,95

Overall Accuracy	0,92	
Rate of agreement	0,25	
Kappa	0,90	

The overall accuracy and kappa indexes are quite similar and quite high. The resulting classification has thus a good level of semantic and geometric qualities, even if a visual inspection is always useful.

Table 9.13. *Validation of the classification*

9.4. Bibliography

[BAR 04] DARDAT J., BIORET F., BOTINEAU M. *et al.*, *Prodrome des végétations de France*, Publications scientifiques du MNHN, Paris, 2004.

[BEN 04] BENZ, U.C., HOFMANN P., WILLHAUCK G, *et al.*, "Multi-resolution, object-oriented fuzzy analysis of remote sensing data for GIS-ready information", *ISPRS Journal of Photogrammetry and Remote Sensing*, vol. 58, no. 3, pp. 239–258, 2004.

[BLA 10] BLASCHKE T., "Object based image analysis for remote sensing", *ISPRS Journal of Photogrammetry and Remote Sensing*, vol. 65, no. 1, pp. 2–16, 2010.

[BOC 03] BOCK M. "Remote sensing and GIS-based techniques for the classification and monitoring of biotopes: case examples for a wet grass-and moor land area in Northern Germany", *Journal for Nature Conservation*, vol. 11, no. 3, pp. 145–155, 2003.

[CAN 14] CANTELAUBE P., CARLES M., "Le registre parcellaire graphique: des données géographiques pour décrire la couverture du sol agricole", *Le Cahier des Techniques de l'INRA* (N Spécial GéoExpé), pp. 58–64, 2014.

[COM 02] COMANICIU D., MEER P., "Mean shift: a robust approach toward feature space analysis", *IEEE Transactions on Pattern Analysis and Machine Intelligence*, vol. 24, no. 5, pp. 603–619, May 2002.

[COR 04] CORBANE C., BAGHDADI N., HOSFORD S. *et al.*, "Application d'une méthode de classification orientée objet pour la cartographie de l'occupation du sol: résultats sur ASTER et Landsat ETM", *Revue Française de Photogrammétrie et de Télédétection*, vol. 175, pp. 13–26, 2004.

[EEA 07] EEA, CLC 2006 technical guidelines, EEA Technical Report, 2007.

[GUO 07] GUO Q., KELLY M., GONG P, *et al.*, "An object-based classification approach in mapping tree mortality using high spatial resolution imagery", *GIScience & Remote Sensing*, vol. 44, no. 1, pp. 24–47, 2007.

[HOO 96] HOOVER A., JEAN-BAPTISTE G., JIANG X. *et al.*, "An experimental comparison of range image segmentation algorithms", *IEEE Transactions on Pattern Analysis and Machine Intelligence*, vol. 18, no. 7, pp. 673–689, 1996.

[HUA 02] HUANG C., DAVIS L.S., TOWNSHEND J.R.G., "An assessment of support vector machines for land cover classification", *International Journal of Remote Sensing*, vol. 23, no. 4, pp. 725–749, 2002.

[IFN 16] IFN, La cartographie forestière version 2, Technical report, p. 51, 2016.

[ING 09] INGLADA J., CHRISTOPHE E., "The Orfeo Toolbox remote sensing image processing software", *Geoscience and Remote Sensing Symposium, 2009 IEEE International* (IGARSS 2009) , IEEE, vol. 4, pp. IV-733, 2009.

[KPA 14] KPALMA K., EL-MEZOUAR M.C., TALEB N., "Recent trends in satellite image pan-sharpening techniques", *1st International Conference on Electrical, Electronic and Computing Engineering*, 2014.

[LAL 04] LALIBERTE A.S., RANGO A., HAVSTAD K.M, *et al.*, "Object-oriented image analysis for mapping shrub encroachment from 1937 to 2003 in southern new mexico", *Remote Sensing of Environment*, vol. 93, nos. 1–2, pp. 198–210, 2004.

[LAN 09] LANG S., SCHÖPFER E., LANGANKE T., "Combined object-based classification and manual interpretation – synergies for a quantitative assessment of parcels and biotopes", *Geocarto International*, vol. 24, no. 2, pp. 99–114, 2009.

[LEE 03] LEEMANS R., DE GROOT R.S., *Millenium Ecosystem Assessment: Ecosystems and Human Well-Being: A Framework for Assessment*, Island Press, Washington, DC, 2003.

[LUC 07] LUCAS R., ROWLANDS A., BROWN A. *et al.*, "Rule-based classification of multi-temporal satellite imagery for habitat and agricultural land cover mapping", *ISPRS Journal of Photogrammetry and Remote Sensing*, vol. 62, no. 3, pp. 165–185, 2007.

[MIK 15] MIKEŠ S., HAINDL M., SCARPA G. *et al.*, "Benchmarking of remote sensing segmentation methods", *IEEE Journal of Selected Topics in Applied Earth Observations and Remote Sensing*, vol. 8, no. 5, pp. 2240–2248, 2015.

[PAL 05] PAL M., "Random forest classifier for remote sensing classification", *International Journal of Remote Sensing*, vol. 26, no. 1, pp. 217–222, 2005.

[TRI 17] TRIMBLE, "eCognition Software", available at http://www.ecognition.com, 2017.

[WEI 08] WEINKE E., LANG S., PREINER M., "Strategies for semi-automated habitat delineation and spatial change assessment in an alpine environment", in BLASCHKE T., LANG S., HAY G. (eds), *Object-Based Image Analysis – Spatial Concepts for Knowledge-Driven Remote Sensing Applications*, Springer, Berlin, 2008.

[YU 06] YU Q., GONG P., CLINTON N. *et al.*, "Object-based detailed vegetation classification with airborne high spatial resolution remote sensing imagery", *Photogrammetric Engineering and Remote Sensing*, vol. 72, no. 7, p. 799, 2006.

List of Authors

Samuel ALLEAUME
IRSTEA/University of Montpellier
TETIS
Montpellier
France

Nicolas BAGHDADI
IRSTEA/University of Montpellier
TETIS
Montpellier
France

Hassan BAZZI
IRSTEA/University of Montpellier
France

Agnès BÉGUÉ
CIRAD/University of Montpellier
TETIS
Montpellier
France

Beatriz BELLÓN
CIRAD/University of Montpellier
TETIS
Montpellier
France

Mar BISQUERT
Universidad de Castilla-La Mancha
Cuenca
Spain

Luca CONGEDO
Sapienza University
Dept. of Architecture and Design
Rome
Italy

Mohammad EL HAJJ
IRSTEA/University of Montpellier
TETIS
Montpellier
France

Pierre-Louis FRISON
IGN-ENSG, University of Paris-Est
LaSTIG
Marne-la-Vallée
France

Raffaele GAETANO
CIRAD/University of Montpellier
TETIS
Montpellier
France

Jean-Marc GILLIOT
AgroParisTech
Grignon
France

Nicolas KARASIAK
INP Toulouse
DYNAFOR
Toulouse
France

Marc LANG
IRSTEA/University of Montpellier
TETIS
Montpellier
France

Cédric LARDEUX
ONF International
LaSTIG
Nogent-Sur-Marne
France

Sylvio LAVENTURE
IRSTEA/University of Montpellier
Montpellier
France

Camille LE PRIOL
AgroParisTech
Grignon
France

Louise LEROUX
CIRAD/University of Montpellier
AIDA
Montpellier
France

Clément MALLET
IGN-ENSG/University of Paris-Est
MATIS
Saint-Mandé
France

Philippe MARTIN
AgroParisTech
Grignon
France

Kenji OSE
IRSTEA/University of Montpellier
TESTIS
Montpellier
France

Pauline PERBET
Parc Amazonien de Guyane
Remire-Montjoly
French Guiana

Juan Manuel SANCHEZ
Universidad de Castilla-La Mancha
Cuenca
Spain

Vincent THIERION
INRA
CESBIO
Toulouse
France

Emmanuelle VAUDOUR
AgroParisTech
Grignon
France

Mehrez ZRIBI
CNRS
CESBIO
Toulouse
France

Index

Scientific Committee

Nicolas BAGHDADI, IRSTEA, University of Montpellier, France

Mehrez ZRIBI, CNRS, France

Clément MALLET, IGN-ENSG, University of Paris-Est, France

Mohammad EL HAJJ, IRSTEA, University of Montpellier, France

Kenji OSE, IRSTEA, University of Montpellier, France

Hassan BAZZI, IRSTEA, University of Montpellier, France

Cécile CAZALS, IGN-ENSG, University of Paris-Est, France

Clément DECHESNE, IGN-ENSG, University of Paris-Est, France

Simon BAILLY, IGN-ENSG, University of Paris-Est, France

Marc POUPÉE, IGN-ENSG, University of Paris-Est, France

Sébastien GIORDANO, IGN-ENSG, University of Paris-Est, France

Tristan POSTADJIAN, IGN-ENSG, University of Paris-Est, France

Other titles from

in

Earth Systems – Environmental Sciences

2017

QUEVAUVILLER Philippe, CIAVOLA Paolo, GARNIER Emmanuel
Management of the Effects of Coastal Storms: Policy, Scientific and Historical Perspectives

2016

BLAIN Stéphane, TAGLIABUE Alessandro
Iron Cycle in Oceans

BRAUDEAU Erik, ASSI Amjad, MOHTAR Rabi
Hydrostructural Pedology

MONACO André, PROUZET Patrick
Seas and Oceans Set
Volume 9 – Tools for Oceanography and Ecosystemic Modeling

QUÉGUINER Bernard
The Biogeochemical Cycle of Silicon in the Ocean

QUEVAUVILLER Philippe
Marine Chemical Monitoring

2015

HERBIN Hervé, DUBUISSON Philippe
Infrared Observation of Earth's Atmosphere

MERCIER-LAURENT Eunika
The Innovation Biosphere

MONACO André, PROUZET Patrick
Seas and Oceans Set
Volume 7 – Marine Ecosystems: Diversity and Functions
Volume 8 – Governance of Seas and Oceans

2014

DE LARMINAT Philippe
Climate Change: Identification and Projections

MONACO André, PROUZET Patrick
Seas and Oceans Set
Volume 1 – Ocean in the Earth System
Volume 2 – The Land-Sea Interactions
Volume 3 – Ecosystem Sustainability and Global Change
Volume 4 – Vulnerability of Coastal Ecosystems and Adaptation
Volume 5 – Development of Marine Resources
Volume 6 – Value and Economy of Marine Resources

MUTTIN Frédéric
Marine Coastal and Water Pollutions: Oil Spill Studies

2013

LÉCUYER Christophe
Water on Earth: Physicochemical and Biological Properties

LEGCHENKO Anatoly
Magnetic Resonance Imaging for Groundwater

2012

LE MENN Marc
Instrumentation and Metrology in Oceanography

PAVÉ Alain
Modeling of Living Systems: From Cell to Ecosystem

2010

TANGUY Jean-Michel
Environmental Hydraulics Series
Volume 1 – Physical Processes and Measurement Devices
Volume 2 – Mathematical Models
Volume 3 – Numerical Methods
Volume 4 – Practical Applications in Engineering
Volume 5 – Modeling Software

Printed and bound by CPI Group (UK) Ltd, Croydon, CR0 4YY

27/10/2024

14580733-0004